161
Advances in Polymer Science

Springer

Berlin
Heidelberg
New York
Hong Kong
London
Milan
Paris
Tokyo

Polymers for Photonics Applications II

Nonlinear Optical, Photorefractive and Two-Photon Absorption Polymers

Volume Editor: K.-S. Lee

With contributions by S.-J. Chung, G. S. He,
A. K-Y. Jen, F. Kajzar, K.-S. Kim, B. Kippelen,
K.-S. Lee, T.-C. Lin, N. Peyghambarian, P. N. Prasad,
H. E. Pudavar, J. Swiatkiewicz, X. Wang

 Springer

This series presents critical reviews of the present and future trends in polymer and biopolymer science including chemistry, physical chemistry, physics and materials science. It is addressed to all scientists at universities and in industry who wish to keep abreast of advances in the topics covered.

As a rule, contributions are specially commissioned. The editors and publishers will, however, always be pleased to receive suggestions and supplementary information. Papers are accepted for „Advances in Polymer Science" in English.

In references Advances in Polymer Science is abbreviated Adv Polym Sci and is cited as a journal.

Springer APS home page: http://link.springer.de/series/aps/ or
http://link.springer-ny.com/series/aps/
Springer-Verlag home page: http://www.springer.de

ISSN 0065-3195
ISBN 3-540-43157-8
Springer-Verlag Berlin Heidelberg New York

Library of Congress Catalog Card Number 61642

Springer-Verlag Berlin Heidelberg New York
a member of BertelsmannSpringer Science+Business Media GmbH
http://www.springer.de

© Springer-Verlag Berlin Heidelberg 2003
Printed in Germany

Typesetting: Data conversion by MEDIO, Berlin
Cover: MEDIO, Berlin
Printed on acid-free paper   02/3020wei - 5 4 3 2 1 0   ac

6-27-03
ac

Volume Editor

Prof. Kwang-Sup Lee
Dept. of Polymer Science & Engineering
Hannam University
133 Ojung-Dong
Taejon 300-791, Korea
E-mail: kslee@mail.hannam.ac.kr

Editorial Board

Preface

The future of information technology requires ultra-high speed processing and large data storage capacity. Since the electronics technology using semi-conductors and inorganic materials is about to reach its limits, much current research is focused on utilizing much faster photons than electrons, namely photonics. To achieve any significant effect on the actual use of the science of photonics, developments of more efficient photonics materials, better optical property evaluations, manufacture of devices for system applications, etc. are the subjects which need to be explored. In particular, the development of photonics materials stands in the forefront of research as this constitutes the most pertinent factor with regard to the development of ultra-high speed and large capacity information processing. In this respect, there has been continuous research on photo-responsive materials through molecular structure design and architecture and the results so far are very promising as functions and performances are beginning to realize their high expectations.

The two special volumes "Polymers for Photonics Applications" give authoritative and critical reviews on up-to-date activities in various fields of photonic polymers including their promising applications. Seven articles have been contributed by internationally recognized and they deal with, polymers for second- and third-order nonlinear optics, quadratic parametric interactions in polymer waveguides, electroluminescent polymers as light sources, photoreflective polymers for holographic information storage, and highly efficient two-photon absorbing organics and polymers. This review should provide individuals working in the field of photonics polymers with invaluable scientific knowledge on the state of the art while giving directions for future research to those deeply interested.

Finally, I would certainly like to give my heartfelt thanks to the contributors to this two volume special issue and although the publication has been delayed, I am very grateful that this issue can be dedicated to Professor Gerhard Wegner in celebration of his 60th birthday.

Taejon, December 2002   Kwang-Sup Lee

Advances in Polymer Science
Available Electronically

For all customers with a standing order for Advances in Polymer Science we offer the electronic form via LINK free of charge. Please contact your librarian who can receive a password for free access to the full articles. By registration at:

http://link.springer.de/series/aps/reg_form.htm

If you do not have a standing order you can nevertheless browse through the table of contents of the volumes and the abstracts of each article at:

http://link.springer.de/series/aps/
http://link.springer-ny.com/series/aps/

There you will find also information about the

– Editorial Board
– Aims and Scope
– Instructions for Authors
– Sample Contribution

Contents

Polymeric Materials and Their Orientation Techniques for
Second-Order Nonlinear Optics
F. Kajzar, K.-S. Lee, A.K-Y. Jen 1

Photorefractive Polymers and their Applications
B. Kippelen, N. Peyghambarian 87

Organics and Polymers with High Two-Photon Activities
and their Applications
T.-C. Lin, S.-J. Chung, K.-S. Kim, X. Wang, G.-S. He, J. Swiatkiewicz,
H. E. Pudavar, P. N. Prasad 157

Author Index 195

Subject Index 209

Contents of Volume 158

Polymer for Photonics Applications I
Nonlinear Optical and Electroluminescence Polymers

Volume Editor: K.-S. Lee

Nonlinear Optical Polymeric Materials: From Chromophere Design to Commercial Applications
L. Dalton

Quadratic Parametric Interactions in Organic Waveguides
M. Canva, G. J. Stegeman

Molecular Design for Third-Order Nonlinear Optics
U. Gubler, C. Bosshard

Light-Emitting Characteristics of Conjugated Polymers
H.-K. Shim, J.-I. Jin

Polymeric Materials and their Orientation Techniques for Second-Order Nonlinear Optics

Francois Kajzar[1] · Kwang-Sup Lee[2] · Alex K-Y. Jen[3]

[1] LETI-CEA (Technologies Avancées), DEIN/SPE, CE de Saclay, 91191 Gif-sur, Yvette Cedex, France
[2] Department of Polymer Science and Engineering, Institute of Hybrid Materials for Information and Biotechnology, Hannam University, Daejeon 306-791, Korea
[3] Department of Materials Science and Engineering, University of Washington, Seattle, WA98195–2120, USA
[1] E-mail: kajzar@ortolan.cea.fr, [2] E-mail: kslee@mail.hannam.ac.kr
[3] E-mail: ajen@u.washington.edu

This review article describes recent developments in the field of second-order nonlinear optical (NLO) polymers with specific focus on their characterization methods, materials synthesis, chromophore orientation techniques, and device applications. For the characterization techniques of NLO properties of organics and polymeric materials, electric field-induced second harmonic generation, hyper-Rayleigh scattering, electro-optic coefficient measurement, etc. are discussed. The significant progress obtained from the authors' investigations, resulted in various types of polymeric materials including dendrimers and organic-inorganic hybrids with specific structures of academic and technological significance which are presented here. To produce highly efficient macroscopic nonlinearity in NLO polymeric systems, several chromophore orientation techniques such as static field poling, photoassisted poling, all optical poling, contact and corona poling are also demonstrated. In addition, the prospects for practical applications of the NLO polymers in information technology are reviewed.

Keywords: Second-order nonlinear optics, Polymers, Dendrimers, Poling, Orientation techniques, Relaxation dynamics, Optical devices

1	**Introduction**	5
2	**Second-Order NLO Properties Characterization Techniques**	6
2.1	Molecular Level	6
2.1.1	Electric Field-Induced Second Harmonic Generation	7
2.1.2	Hyper-Rayleigh Scattering	8
2.2	Macroscopic Level	9
2.2.1	Second Harmonic Generation	9
2.2.2	Linear Electro-Optic Effect	10
3	**Polymeric Materials for Second-Order Nonlinear Optics**	13
3.1	Guest-Host-Type Polymers	13
3.2	Side-Chain NLO Polymers	17
3.2.1	Polyurethane-Based NLO Polymers	17
3.2.2	Polyimide-Based NLO Polymers	21

Advances in Polymer Science, Vol. 161
© Springer-Verlag Berlin Heidelberg 2003

3.3   Crosslinked NLO Polymers.   26
3.4   Hyperbranched and Dendritic Polymers   28
3.4.1   Spontaneous, Noncentrosymmetric Organization
        of NLO Chromophore in Dendritic Structures
        (NLO Dendritic Effect at the Molecular Level)   28
3.4.2   Enhancement of Poling Efficiency in NLO Dendrimers
        (NLO Dendritic Effect at the Macroscopic Level).   32
3.4.2.1 Single-Dendron-Modified NLO Chromophores   32
3.4.2.2 Multiple-Dendron-Modified NLO Chromophores.   34
3.4.2.3 Multiple-Chromophore-Containing Crosslinkable NLO Dendrimers   37
3.5   Organic-Inorganic Hybrid Materials   44

4   **Chromophore Orientation Techniques**   50

4.1   Static Field Poling.   50
4.1.1   Establishment of the Axial Order   53
4.1.2   Build-Up of the Polar Order   57
4.2   Photoassisted Poling   60
4.3   All Optical Poling   62

5   **Statistical Orientation Models**   64

6   **Relaxation Processes**.   70

7   **Light-Induced Depoling**.   71

8   **Applications of Second-Order NLO Polymers**.   73

8.1   Frequency Doubling.   73
8.1.1   Cerenkov-Type SHG Generation.   74
8.1.2   Quasi-Phase Matching in Periodically Poled Polymer Films.   74
8.1.3   Counter-Propagating Beams Second Harmonic Generation   75
8.1.4   Modal Phase Matching   76
8.2   Frequency Tuning   76
8.3   Electro-Optic Modulation   77
8.4   Ultrashort Electric Pulse Generation   79

9   **Concluding Remarks**.   79

References   80

List of Abbreviations and Symbols

APC	amorphous polycarbonate
α	molecular polarizability
a.u.	arbitrary unit
β	molecular first hyperpolarizability
BOA	bond-order alternation
BSB	Babinet-Soleil-Bravais
c	light speed
d_{33}	nonlinear optical coefficient
DAST	4'-dimethylamino-N-methyl-4-stilbazolium tosylate
DC	direct current
DEAD	diethyl azodicarboxylate
$\Delta\varepsilon$	dielectric constant dispersion
DFG	difference frequency generation
DMAc	dimethylacetamide
DMF	dimethylformamide
DMSO	dimethylsulfoxide
D-π-A	donor–π center–acceptor
DR 1	disperse red 1
DSC	differential scanning calorimetry
E	electric poling field strength
ε_0	low frequency dielectric constant
$\varepsilon(2\omega)$	molar extinction coefficient of polymer at 2ω frequency
EFISH	electric field-induced second harmonic generation
esu	electrostatic unit
F	local field factor
FTC	tricyanofuran
G(θ)	Gibbs-Boltzmann distribution function
GHz	gigahertz (10^9 Hz)
γ	molecular second hyperpolarizability
h	Planck's constant
HLS	harmonic light scattering
HRS	hyper-Rayleigh scattering
Hz	hertz (1 Hz = s^{-1})
I_ω	fundamental beam intensity
ITO	indium tin oxide
K_i	expansion coefficients
KWW	Kohlrausch-Williams-Watt
μ	dipole moment
μ_0	permanent dipole moment
LB	Langmuir-Blodgett

MHz	megahertz (10^6 Hz)
MV	megavolt (10^6 V)
M_w	Molecular weight
n	refractive index
N	chromophore number density in polymer matrix
NLO	nonlinear optical
NMP	N-methylpyrrolidone
NPP	N-(4-nitrophenyl)-(S)-prolinol
OPO	optical parametric oscillator
P	macroscopic polarization
PC	polycarbonate
PEI	polyetherimide
Ph-TCBD	2-phenyl-tetracyanobutadienyl
PI	polyimide
PMMA	poly(methyl methacryate)
PPIF	poly[(phenyl isocyanate)-co-formaldehyde]
PQ	polyquinoline
PS	polystyrene
PU	polyurethane
QPM	quasi-phase matched
r_{33}	electro-optic (E-O) coefficient in the direction of the applied electric field
λ	wavelength
RT	room temperature
SG	sol-gel
SHG	second harmonic generation
TCN	tricyano
T_g	glass transition temperature
THF	tetrahydrofuran
THG	third harmonic generation
UV	ultraviolet
V_m	amplitude of modulation voltage
V_p	half-wave voltage
$\chi^{(1)}$	linear optical susceptibility
$\chi^{(2)}$	second-order nonlinear optical susceptibility
$\chi^{(3)}$	third-order nonlinear optical susceptibility
ω	angular frequency

1
Introduction

Since the late sixties organic molecules have attracted an increasing amount of interest due to their potential applications in nonlinear optical (NLO) devices, and in particular in second-order nonlinear optics. This interest is motivated not only by the large NLO response, but also by the versatility, ease of processing, and possibility of tailoring the physicochemical properties by the molecular engineering approach. It is well known that the organic molecules, due to their enhanced hyperpolarizability, exhibit large, fast, electronic in origin, second-order nonlinear NLO response and may be processed into good optical quality thin films and single crystals.

The nonlinear optical properties arise from the ability of molecules and atoms to change, in a nonlinear way, their polarization under the external forcing field (E). At the molecular level, the induced variation of the dipole moment (μ) may be developed into the electric field power series giving variation of molecule dipole moment:

$$\mu - \mu_0 = K_1 \alpha : E + K_2 \beta : EE + K_3 \gamma : EEE + \cdots \cdots \tag{1}$$

where the expansion coefficients are three-dimensional tensors describing molecular polarizability (α), molecular first (β) and second hyperpolarizability (γ). The expansion coefficents (K_is) in Eq. (1) depend on units and conventions used for the Fourier transform of the electric field as well as on whether the permutation factors arising from the degeneracy of electric field are, or not, taken explicitly into account. Hereafter, we use a convention in which these factors are taken explicitly into account and a coefficient of 1/2 is used in the Fourier transform of the electric field [1]. The electric fields appearing in Eq. (1) are local fields, which differ from the applied external fields due to the screening by an internal field.

Similar development obeys also in the laboratory reference frame. Again, under the external forcing field the medium polarization (P) can be expanded, within the dipolar approximation in the power series of the external forcing field E giving

$$P = P_0 + K_1 \chi^{(1)} : E + K_2 \chi^{(2)} : EE + K_3 \chi^{(3)} : EEE + \cdots \tag{2}$$

where the development coefficients $\chi^{(n)}$ are three-dimensional ($n + 1$) rank tensors describing linear ($\chi^{(1)}$) and NLO response ($\chi^{(2)}, \chi^{(3)}, ...$) of material in the laboratory reference frame. The factors K_i have the same meaning as in Eq. (1).

As it is seen from Eqs. (1) and (2) the values of molecular hyperpolarizabilities and macroscopic susceptibilities, within a given system of units, depend directly on the conventions used. Therefore, it is important when comparing data coming from different determinations or with theoretical calculations to ensure what kind of conventions were used.

For centrosymmetric materials, all odd rank tensors $\chi^{(2)}, \chi^{(4)}, \ldots \equiv 0$ and for centrosymmetric molecules, all odd rank molecular hyperpolarizabilities $\mu, \beta, \ldots \equiv 0$. Thus, the search for new materials for second-order NLO applications consists on the synthesis of highly efficient, noncentrosymmetric molecules such as charge-transfer molecules (or more recently octupolar molecule [2, 3]) and assembling these molecules into noncentrosymmetric bulk materials: polymer thin films or single crystals, depending on targeted application. The former approach is called molecular engineering while the latter is called material engineering. Both aspects will be reviewed and discussed in this paper, along with a short description of characterization techniques and device applications.

2
Second-Order NLO Properties Characterization Techniques

Several techniques were developed for the characterization of nonlinear optical properties of molecules and of bulk materials. The microscopic hyperpolarizabilities are obtained from the macroscopic ones by using corresponding relationships between them. This is not always straightforward because of the complicated character of the dependence between these quantities (cf. Chapter 3). However, due to the molecule and bulk material symmetry and the Kleinman relationships, the number of the nonvanishing β and $\chi^{(2)}$ tensor components (27 in general) is strongly reduced. In particular, of the charge-transfer molecules considered here, the β_{zzz} component in the charge-transfer direction is enhanced and the others can be neglected.

2.1
Molecular Level

Basically, three techniques are used for determination of the first hyperpolarizability β tensor components:
(i)   Electric field-induced second harmonic generation
(ii) Hyper-Raleigh scattering (HRS)
(iii) Solvatochromism

We will shortly describe only the first two techniques. Solvatochromism [4] is a very crude technique and gives approximate values of β only. The values depend on the solvent used (interaction between solvent and solute molecules), arbitrary parameters such as cavity radius, and on the validity of the two-level model [5].

2.1.1
Electric Field-Induced Second Harmonic Generation

The first common method for molecular first hyperpolarizability determination is the electric field-induced second harmonic generation (EFISH) technique in solution [6–10]. This technique can be applied only to dipolar molecules. Under an applied external electric field, molecules in solution orient approximately in the direction of the field giving rise to second harmonic generation. The measured third-order nonlinear optical susceptibility is given by the following expression:

$$\chi^{(3)}(-2\omega;\omega,\omega,0) = NF\left[\gamma(-2\omega;\omega,\omega,0) + \frac{\mu_z\beta_z}{5kT}\right] \tag{3}$$

where z is the direction of the molecule dipole moment (usually the direction of the charge transfer axis), N is the number density of solute molecules and the second hyperpolarizability is given by

$$\gamma(-2\omega;\omega,\omega,0) = \frac{1}{5}\left[\gamma_{xxxx} + \gamma_{yyyy} + \gamma_{zzzz} + 2\left(\gamma_{xxyy} + \gamma_{xxzz} + \gamma_{yyzz}\right)\right] \tag{4}$$

The vector part of the first hyperpolarizability tensor is given by

$$\beta_z(-2\omega;\omega,\omega) = \beta_{zzz} + \beta_{zxx} + \beta_{zyy} \tag{5}$$

Thus, in this technique one measures only the vector part of β. As already mentioned, in the case of charge-transfer molecules under consideration here, the β_{zzz} component is strongly enhanced and the other components intervening in Eq. (5) are negligible.

The local field factor F, intervening in Eq. (3) takes account of the already-mentioned screening effects and is given by

$$F = (f_\omega)^2 f_{2\omega} f_0 \tag{6}$$

where

$$f_{\omega(2\omega)} = \frac{\left(n_{\omega(2\omega)}\right)^2 + 2}{3} \tag{7}$$

and

$$f_0 = \frac{\varepsilon_0\left[\left(n_{\omega(2\omega)}\right)^2 + 2\right]}{\left(n_{\omega(2\omega)}\right)^2 + 2\varepsilon_0} \tag{8}$$

where ε_0 is the low frequency dielectric constant.

The second hyperpolarizability $\gamma(-2\omega;\omega,\omega,0)$ on the right-hand side of Eq. (3) is usually neglected with respect to the orientational β term. This is almost justified in the case of small, weakly conjugated molecules, as $\gamma(-2\omega;\omega,\omega,0)$ increases much stronger with conjugation length than $\beta(-2\omega;\omega,\omega)$. This assumption is not satisfied in the case of more conjugated molecules. In that case, the second hyperpolarizability $\gamma(-2\omega;\omega,\omega,0)$ has to be determined by other techniques, for example, third harmonic generation (THG). As a matter of fact, the THG technique gives the $\gamma(-3\omega;\omega,\omega,\omega)$ hyperpolarizability which differs from that intervening in Eq. (3) because of conventions and different dispersion laws. A correct approach consists of the measurement of the $\gamma(-3\omega;\omega,\omega,\omega)$ spectrum. Then, by applying the essential state model, one can determine different dipolar transition moments and differences of dipolar moments between fundamental and excited states in the case of noncentrosymmetric molecules. This can then be used to calculate the $\gamma(-2\omega;\omega,\omega,0)$ term at a given (measurement) frequency [5,11,12]. More details on the EFISH technique can be found in a critical review by Kajzar [9].

2.1.2
Hyper-Rayleigh Scattering

Hyper-Rayleigh scattering (HRS) or harmonic light scattering (HLS) [13–15] is a relatively simple technique allowing determination of different components of first hyperpolarizability β without molecular orientation. Thus, the measurements may be performed on molecules with no permanent dipole moment such as the already-mentioned octupolar molecules. The measurements are done in solution, and the scattered light intensity in the y-direction for the z-polarized incoming beam is given by [15]:

$$I_z^{2\omega} = g\left[N_s\left\langle\beta_{zzz}^2\right\rangle_s + N_p\left\langle\beta_{zzz}^2\right\rangle_p\right]e^{-\varepsilon(2\omega)lN_p}I_\omega^2 \tag{9}$$

$$I_x^{2\omega} = g\left[N_s\left\langle\beta_{zxx}^2\right\rangle_s + N_p\left\langle\beta_{zxx}^2\right\rangle_p\right]e^{-\varepsilon(2\omega)lN_p}I_\omega^2 \tag{10}$$

where N_s and N_p are number densities of the solvent (s) and solute (p), respectively, $\varepsilon(2\omega)$ is the molar extinction coefficient of polymer at 2ω frequency and l is an effective optical path. The factor g in Eqs. (9) and (10) takes account of local fields and geometrical factors, and the averages are given by:

$$\left\langle\beta_{zzz}^2\right\rangle = \frac{1}{7}\sum_i\beta_{iii}^2 + \frac{6}{35}\sum_{i\neq j}\beta_{iii}\beta_{ijj} + \frac{9}{35}\sum_{i\neq j}\beta_{ijj}^2 + \frac{6}{35}\sum_{i,j,k,cyclic}\beta_{iij}\beta_{jkk} + \frac{12}{35}\beta_{ijk}^2 \tag{11}$$

$$\langle \beta_{zxx}^2 \rangle = \frac{1}{35} \sum_i \beta_{iii}^2 - \frac{2}{105} \sum_{i \neq j} \beta_{iii} \beta_{ijj} + \frac{11}{105} \sum_{i \neq j} \beta_{ijj}^2 - \frac{2}{105} \sum_{i,j,k,cyclic} \beta_{iij} \beta_{jkk} + \frac{8}{35} \beta_{ijk}^2 \quad (12)$$

Although, *a priori*, there are a large number of β tensor components intervening in Eqs. (9)–(12) for the limited amount of experimental data, their number is usually limited by molecular symmetry and Kleinmen conditions. Moreover, some of the components can be neglected with respect to others.

The β tensor components are obtained from the slope of the scattered light intensity at harmonic frequency versus the square of the fundamental beam intensity, measured for different polarization configurations. The value of the g parameter intervening in Eqs. (9)–(11) is obtained by calibration with a solvent of known β value. These HRS measurements have to be done very carefully, as other processes such as two-photon-induced fluorescence, Raman, and higher order NLO processes can contribute to the measured HRS signal. Actually, the use of the time-resolved technique with femtosecond pulses allows for the separation of photons coming from the harmonic generation of the others.

2.2
Macroscopic Level

Principally two techniques are used to measure the second-order nonlinear optical susceptibilities. These are:
(i)   Second harmonic generation
(ii)   Linear electro-optic (Pockels) effect

Both can be done on single crystals and on oriented thin films. The former yields $\chi^{(2)}(-2\omega;\omega,\omega)$ susceptibility, while the latter $\chi^{(2)}(-\omega;\omega,0)$. Depending of the relative orientation of the incident and output fields with respect to the crystal (or thin film) symmetry axes, different components of both susceptibilities can be obtained. The SHG technique is also often used to study the kinetics of orientation and relaxation processed by in situ measurements [16].

2.2.1
Second Harmonic Generation

The second harmonic generation is a coherent technique giving the fast, electronic in origin, second-order NLO susceptibility $\chi^{(2)}(-2\omega;\omega,\omega)$ at a given, measurement frequency ω. Here, we limit the discussion to poled films, with ∞ mm symmetry, which exhibit two nonzero $\chi^{(2)}$ tensor components: the diagonal $\chi^{(2)}_{zzz}(-2\omega;\omega,\omega)$ and the off diagonal $\chi^{(2)}_{xzz}(-2\omega;\omega,\omega)$, where Z is the poling (preferential orientation) direction. Usually, thin films are deposited on one side of substrate only (thin film deposited on both sides is discussed in Swalen and Kajzar [17]). The thin film with substrate is mounted on a goniometer. The SHG intensity is collected as function of the incidence angle θ, rotating thin film with substrate around an axis perpendicular to the beam propagation direction and coinciding with it. For the poled thin film symmetry (poling direction perpendicular to the thin film surface) the harmonic intensity is given by (neglecting multiple reflection effects):

$$I_{2\omega}^{ffh} = \frac{\pi^2}{2\varepsilon_0} \left[\frac{\chi_{ffh}^{(2)}(-2\omega;\omega,\omega)}{\Delta\varepsilon} \right]^2 \left[P_{ffh}(\theta) \right]^2 \left[T_{ffh}(\theta)\left(e^{i\Delta\varphi_1} - 1 \right) \right]^2 (I_\omega)^2 \qquad (13)$$

where T_{ffh}s are transmission and boundary condition factors for a given fundamental (f) and harmonic (h) polarization configuration, I_ω is the fundamental beam intensity and $\Delta\varepsilon = n_\omega^2 - n_{2\omega}^2$ is the dielectric constant dispersion.

As already mentioned, the SHG measurements are done in the fundamental (f) and harmonic (h) polarization configuration, starting with the s-polarized fundamental and p-polarized harmonic beams. In that case the projection factor is given by:

$$P_{ssp} = \chi_{ssp}^{(2)} \sin\left(\theta_{2\omega}^p \right) \qquad (14)$$

By using appropriate calibration (e.g., a quartz single-crystal plate) the off diagonal susceptibility $\chi_{ssp}^{(2)}$ (or $\chi_{sp}^{(2)}$ or $d_{sp} = \frac{1}{2}\chi_{sp}^{(2)}$) can be deduced directly from these measurements. Then, by doing the SHG measurements at p-p fundamental harmonic beam configuration, for which the projection factor is given by:

$$P_{ppp} = \chi_{ppp}^{(2)} \sin\left(\theta_{2\omega}^p \right)\sin^2 \theta_\omega^p + \chi_{ssp}^{(2)}\left(\sin 2\theta_\omega^p \cos\theta_{2\omega}^p + \sin\theta_{2\omega}^p \cos^2 \theta_\omega^p \right) \qquad (15)$$

one can determine the diagonal $\chi_{ppp}^{(2)}$ (or $\chi_{pp}^{(2)}$, or $d_{pp} = \frac{1}{2}\chi_{pp}^{(2)}$) susceptibility tensor component, substituting into Eq. (15) the previously determined $\chi_{ssp}^{(2)}$ term with an appropriate calibration. $\Delta\varphi$ in Eq. (13) is the phase mismatch between fundamental and harmonic beams and T_{ffh} is a global transmission factor [17].

2.2.2
Linear Electro-Optic Effect

The linear electro-optic effect arises from the ability of a material medium to change its refractive index under the action of an external electric field. This variation is proportional to the external field strength

$$\delta n_{ii} = p_{ij}\mathrm{E}_j \qquad (16)$$

where the Pockels tensor components p_{ij} are related to the second-order NLO through the following relationship [10]:

$$p_{ij} = \frac{\chi_{iij}^{(2)}(-\omega;\omega,0)}{n_{ii}} \qquad (17)$$

Often, phenomenologically, one uses the linear electro-optic coefficient r_{ij} which, is also directly related to the electro-optic susceptibility:

$$r_{ij} = \frac{2\chi_{iij}^{(2)}(-\omega;\omega,0)}{n_{ii}^3} \qquad (18)$$

Several techniques have been developed to measure the electro-optic suscepti-bility of single crystals or thin films. All are based on the measurement of variation of the refractive index of a material induced by the applied external field. We will shortly describe the modulation ellipsometry technique [18, 19], which is relative-ly simple in use and can be applied to thin films, provided that some precautions are undertaken [20].

For a multilayer structure (thin film sandwiched between two electrodes) (Fig. 1) in reflection (the second electrode is then opaque), the reflected (similar measurements can be done also in transmission) beam intensity is given by:

$$I_r = I_i \sin^2\left(\Psi_{ps} + \Psi_B\right) \qquad (19)$$

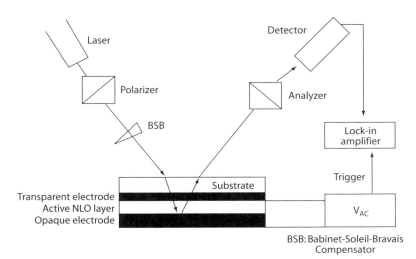

BSB: Babinet-Soleil-Bravais Compensator

Fig. 1. Schematic representation of multilayer structure and experimental set-up for E-O co-efficient measurements by the modulation ellipsometry technique

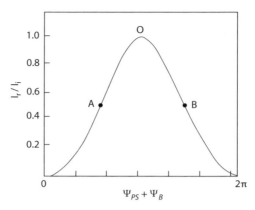

Fig. 2. Variation of the reflected intensity (normalized to the incident intensity) as a function of the phase mismatch induced by the external electric field (Ψ_{PS}), and by BSB (Ψ_B)

where Ψ_{ps} is the phase mismatch inside the film between s and p waves given by:

$$\Psi_{ps} = \frac{4\pi l}{\lambda}\left[\frac{n_0}{n_e}\sqrt{n_e^2 - n_a^2 \sin^2\theta_a} - \sqrt{n_0^2 - n_a^2 \sin^2\theta_a}\right] \qquad (20)$$

where $n_{o(e)}$ is the ordinary (extraordinary) index of refraction, n_a is the refractive index of the surrounding atmosphere, θ_a is incidence angle, and Ψ_B is the phase mismatch between s and p waves induced by the Babinet-Soleil-Bravais (BSB) compensator (for details see [10]), introduced in the optical path and l is the thin film thickness. The BSB is set in such a way as to get the maximum derivative of the transmitted intensity (points A and B in the transmission curve, Fig. 2).

$$\frac{\partial I_r}{\partial E} = \sin\left[2\left(\Psi_{ps} + \Psi_B\right)\right]\frac{\partial \Psi_{ps}}{\partial E} \qquad (21)$$

What happens for the case of $\Psi_{ps} + \Psi_B = \pm\pi/2$? From the derivative $\frac{\partial \Psi_{ps}}{\partial E}$ in Eq. (21) and in the case of small anisotropy in the film ($n_o \approx n_e = n$), one can directly obtain the electro-optic coefficient by measuring the amplitudes of modulation I_m^A and I_m^B at points A and B, respectively, to get the real (r) and imaginary (i) parts [20] of the diagonal r tensor component ($r = r^r + r^i$):

$$r_{33}^r = \frac{I_m^A - I_m^B}{FI_i V_m} \qquad (22)$$

$$r_{33}^i = \frac{I_m^A + I_m^B}{FI_i V_m} \qquad (23)$$

where

$$F = \frac{4\pi}{3\lambda} \frac{n^2 n_a^2 \sin^2 \theta_a}{\sqrt{n^2 - n_a^2 \sin^2 \theta_a}} \tag{24}$$

and V_m is the amplitude of modulation voltage. In deriving Eqs. (22)–(24), for the sake of simplicity, it is assumed that the ratio $r_{33}/r_{13} = 3$, as it is in the case of moderately poled polymer thin films (gas-phase model).

As already mentioned, the reflection modulation technique is relatively simple but for thicker or absorbing thin films it becomes much more complicated due to the multiple reflection effects in the used multilayer structure (Fig. 1) which may lead to erroneous results, if not correctly taken into account [10, 20]. In that case, the measurement of the incidence angle dependence of the modulation intensity is required. Through a correct analysis of experimental data one can get both real and imaginary parts of r, as well as its anisotropy (r_{33} and r_{13} tensor components).

3
Polymeric Materials for Second-Order Nonlinear Optics

It is well known that the second-order NLO properties originate typically from noncentrosymmetric alignment of NLO chromophores in poled polymers. To obtain device-quality materials, three stringent issues must be addressed [21]: (i) design and synthesis of high $\mu\beta$ chromophores and realization of large macroscopic E-O activity in the chromophore-incorporated polymers; (ii) maintenance of long-term temporal stability in the E-O response of the poled materials in addition to their high intrinsic stability toward the environment such as heat, light, oxygen, moisture, and chemicals; (iii) minimization of optical loss from design and processing of materials to fabrication and integration of devices. From the point of viewing the distribution and bonding types of NLO chromophores in polymers, the polymeric materials for second-order NLO can be divided into guest-host type polymers, linear-type (side-chain and main-chain) polymers, crosslinked-type polymers, hyperbranched and dendritic polymers, and organic-inorganic hybrids.

3.1
Guest-Host-Type Polymers

Quantum mechanical analysis based on a simple two-level model [22] and bond-order alternation (BOA) principle exploiting aromaticity [23] have worked surprisingly well in providing useful structure/property relationships for the design of chromophores with ever improving molecular hyperpolarizability. Table 1 provides some representative examples with improved molecular optical nonlinearity developed over the past decade. It has been shown that very large nonlinearities

Table 1. The $\mu\beta$ values for representative NLO chromophores and r_{33} values for their guest-host polymers

NLO chromophore	$\mu\beta$ (10^{-48} esu) at 1907 nm	$\mu\beta/M_{\mathrm{w}}$	NLO chromophore	$\mu\beta$ (10^{-48} esu) at 1907 nm	$\mu\beta/M_{\mathrm{w}}$
1	580 $r_{33} = 13$ pm V^{-1} at 1330 nm (30% wt. in PMMA)	2.1	5	13,500 $r_{33} = 55$ pm V^{-1} at 1330 nm (20% wt. in PC)	27.1
2	6200 $r_{33} = 30$ pm V^{-1} at 1330 nm (25% wt. in Polyimide)	17.3	6	$r_{33} = 65$ pm V^{-1} at 1330 nm (20% wt. in PMMA)	
3	9800 $r_{33} = 45$ pm V^{-1} at 1330 nm (25% wt. in PQ-100)	25.5	7	35,000 $r_{33} > 60$ pm V^{-1} at 1330 nm (30% wt. in PMMA) $V_{\mathrm{p}} = 0.8$ V	45.7
4	$r_{33} = 36$ pm V^{-1} at 1330 nm (25% wt. in PQ-100)		8	$r_{33} = 105$ pm V^{-1} at 1330 nm (17.5% wt. in PMMA)	

can be achieved by combining heterocyclic conjugating units, such as thiophene with tricyanovinyl [24] or [3-(dicyanomethylidene)-2,3-dihydrobenzothiophen-2-ylidene-1,1-dioxide] electron acceptors [25], or by employing extended polyene π-bridged systems with strong multicyano-containing heterocyclic electron acceptors [26,27]. For many years, it had been suggested that E-O polymers could show electro-optic coefficients (r_{33}) much larger than that of the technologically important crystal, lithium niobate (LiNbO$_3$). By using several highly nonlinear molecules such as **5** [25], **6**, **7** [28], and **8** [29] shown in Table 1 in several low-glass transition temperature (T_g) polymers, such as poly(methylmethacryate) (PMMA) and polycarbonate (PC), it was shown that very large E-O coefficients ($r_{33} > 60$ pm V^{-1}, measured at 1.3 μm) were achievable. This is twice the value for lithium niobate (31 pm V^{-1}).

The molecular orientation in the poled polymers is thermodynamically unstable and quickly decays in low-T_g polymers such as PMMA, resulting in a greatly reduced nonlinearity. However, if the T_g of the polymer is roughly 150–200 °C above the ultimate operating temperature, decay of the orientation would be negligible over the device lifetime. Many kinds of NLO chromophores were incorporated as guests in high-T_g polymers, such as polyimide and polyquinoline thin films. After poling, the polymer showed much improved long-term stability at elevated temperatures. This was first demonstrated with a commercial polyimide and NLO chromophore [30], and later with highly nonlinear heteroaromatic chromophores that were developed by Jen et al. using donor-acceptor substituted thiophene-containing conjugated units. The chromophore **2** was stable enough to sustain both imidization (30 min at 200 °C) and poling (10 min at 220 °C) conditions [31]. Through the use of high-T_g polyimide and a physical aging process, this guest-host NLO polymer had long-term stability at elevated temperatures of 120 °C and 150 °C for more than 30 days, with the electro-optic coefficient maintained at 80% (60%) of its initial value.

A similar but more active chromophore **3** (RT-9800) was also incorporated as guest in a rigid-rod, high-temperature polyquinoline (PQ-100) (Scheme 1) [32]. Poling results from the guest-host polyquinoline thin films showed both exceptionally large electro-optic activity and long-term stability at 85 °C. After an initial drop from 45 to 26 pm V^{-1} in the first 100 hours, the electron-optic coefficient remained at 26 pm V^{-1} for more than 2000 hours (Fig. 3).

The experimental value of r_{33} also agrees fairly well with the predicted value of 48 pm V^{-1} that was calculated from $\mu\beta$ by using a two-level model suggested by Katz and Singer et al. [33], after accounting for dispersion effects from both the EFISH and electro-optic measurements. The r_{33} is expressed as:

$$r_{33} = (4/n^4)N(f_\omega)^2(f_0)^2(\mu\beta/5kT)E_p\left[F_1(\omega_0,\omega')/F_2(\omega_0,\omega)\right] \tag{a}$$

where N is the number density of the chromophores, n is the index of refraction, f_o and f_w are the local field factors

$$f_0 = \varepsilon(n^2+2)/(n^2+2\varepsilon) \qquad f\omega = (n^2+2)/3 \tag{b}$$

and F_1 and F_2 are the dispersion factors

$$F_1(\omega_0,\omega') = \omega_0^4 \Big/ \left[\left(\omega_0^2-\omega'^2\right)\left(\omega_0^2-4\omega'^2\right)\right] \tag{c}$$

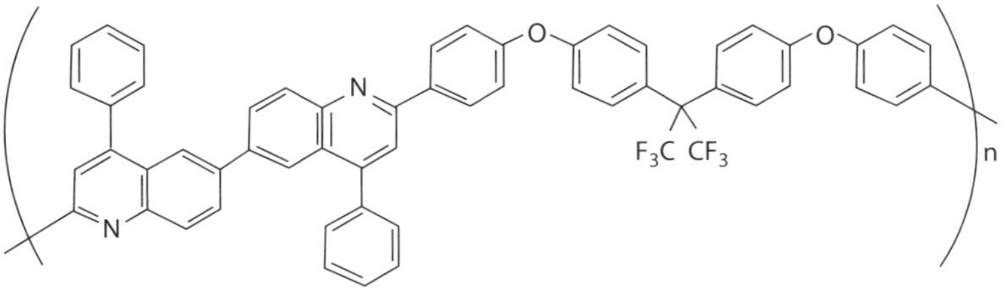

Scheme 1

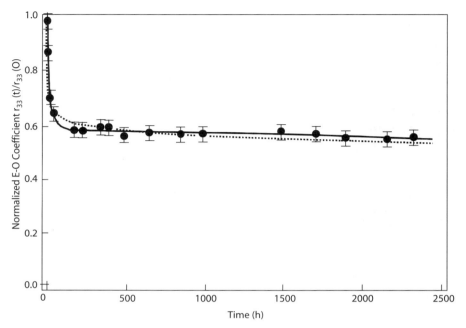

Fig. 3. Normalized E-O coefficients of RT-9800/polyquinoline at 80 °C. The *data points* are the experimental values. The *dotted line* is the "KWW" stretched exponential function fitting and the *solid line* is a biexponential function fitting

and

$$F_2(\omega_0,\omega) = \omega_0^2\left(3\omega_0^2 - \omega^2\right) \Big/ \left[3\left(\omega_0^2 - \omega^2\right)^2\right] \qquad (d)$$

Here ω' is the optical frequency used in the EFISH measurement, ω is the frequency adopted in the electro-optic measurement, and ω_0 is the transition energy between the ground state and first excited state of the chromophore.

In the continuous quest for chromophores that possess both large molecular nonlinearity and good stability, Wu and Jen et al. [34] have developed another series of highly efficient, thermally and chemically stable chromophores such as **4**, based on the 2-phenyltetracyanobutadienyl (Ph-TCBD) group as the electron acceptor. An X-ray single-crystal structure of such a chromophore reveals that the dicyanovinylenyl moiety is linked coplanarly to the donor-substituted aryl segment and forms an efficient push-pull system. In the meantime, the "substituent", α-Ph-dicyanovinyl is twisted out of the main conjugation plane. The three-dimensional shape of the chromophore may help to prevent molecules from stacking up on each other, and thus, reduce chromophore aggregation. This, in turn, improved the poling efficiency and lowered the optical loss caused by light scattering. This chromophore was incorporated as guest (20 wt%) in polyquinoline thin films. After poling, a very high electro-optic coefficient (36 pm V^{-1} at 1.3 μm) was achieved and the value remained at approximately 80% of its original value at 85 °C for more than 1000 hours.

3.2
Side-Chain NLO Polymers

In side-chain NLO polymers, generally, NLO chromophores are incorporated by covalent bonding to various polymer backbones. As compared to guest-host systems, these types of polymers can provide higher chromophore loading density. They also improve temporal and thermal stability of aligned NLO dipoles. Vinyl polymers like polystyrene (PS) [35] and poly(methyl methacrylate) (PMMA) [36–40], and high-Tg polymers such as polyurethanes [41–43], polyimides [44–53], polyamides [54, 55], polyesters [56–58], polyethers [59], and polyquinolines [60–62] are some candidate polymers being considered for NLO chromophore incorporation. Among these, polyurethanes and polyimides are very promising materials as their properties including thermal stability are shown to be excellent. The following discussions are focused on these two polymer systems.

3.2.1
Polyurethane-Based NLO Polymers

Polyurethanes are advantageous as a NLO polymer matrix in that an extensive formation of hydrogen bonding between the urethane linkage would increase the rigidity of the polymer matrix. This fact is attributed to a strong intermolecular interaction that restrains the molecular motion, thereby retarding the relaxation of the aligned NLO dipoles. NLO polyurethanes can be synthesized by the step-growth polymerization reaction of diols, which possess various NLO chromophores, with aromatic diisocyanate such as 2,4-toluene diisocyanate or 4,4'-diisocyanato-3,3'-dimethoxyphenyl (Scheme 2). The resulting polymers are readily soluble in aprotic polar solvents, for example, dimethylformamide (DMF), *N*-methylpyrrolidone (NMP), or cyclohexanone. From the polymer solutions of these solvents, good optical quality films can be cast on glass or quartz substrates by the spin coating method. The physical data and $\chi^{(2)}$ coefficient of some representative NLO polyurethanes are listed in Table 2 [63–69]. Differential scanning calorimetry (DSC) analysis data show relatively high T_g in the temperature range 121–176 °C for the linear polyurethanes. No melting points are detected in these polymers, suggesting that these polyurethanes are amorphous.

The macroscopic second-order susceptibility $\chi^{(2)}$ of these materials is determined by measuring the SHG signal intensity. By optimizing the poling process,

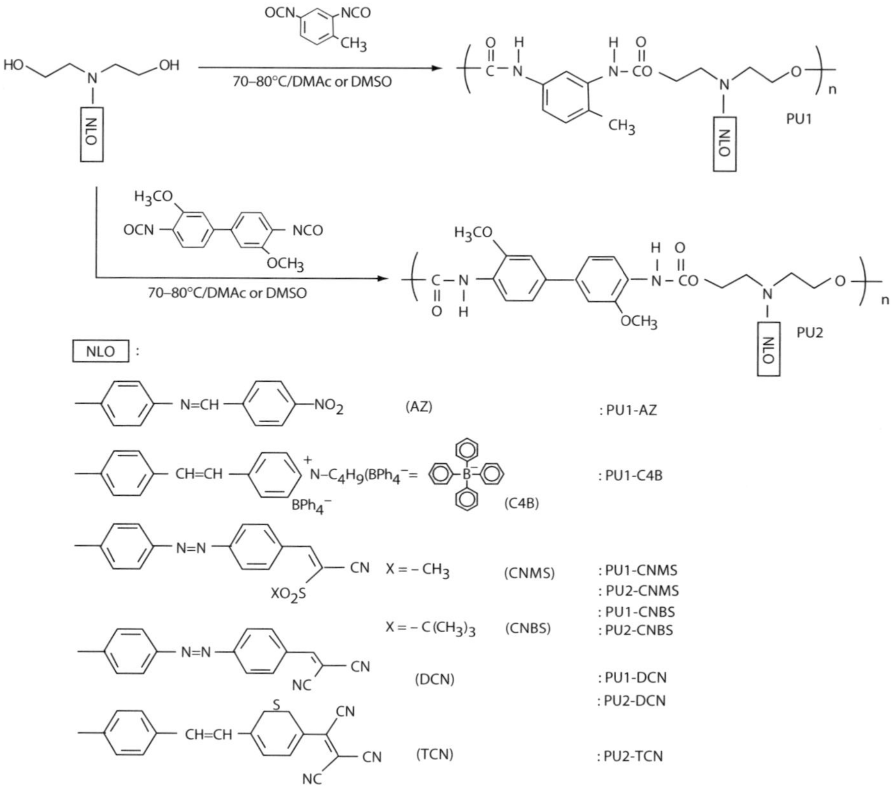

Scheme 2

Table 2. Physical data and $\chi^{(2)}$ coefficients for second-order NLO[a] polyurethanes

Polyurethanes	Glass transition temperature (T_g) (°C)	UV (λ_{max}) (nm)	Refractive index (n) (TM)	$\chi^{(2)}$ (pm V^{-1})	($\times 10^{-7}$ esu)	Reference
PU1-AZ	137	460	1.7013	20	0.5	63, 64
PU1-C4B	121	484	1.7410	128	3.1	65
PU1-CNMS	140	476	1.8607	57	1.4	66
PU2-CNMS	127	490	1.8712	50	1.2	66
PU1-CNBS	153	459	1.7178	57	1.4	67
PU2-CNBS	131	500	1.7208	40	1.0	67
PU1-DCN	159	489	1.7799	136	3.3	68
PU2-DCN	159	496	1.7572	120	2.9	68
PU2-TCN	176	659	1.7800	51	1.2	69

[a] For evaluation of the SHG activity, the Maker Fringe method was used [70]. A Y-cut quartz plate ($d_{11} = 0.81 \times 10^{-10}$ esu) was employed as the reference [71]. A mode-locked Q-switched Nd:YAG laser was used as the fundamental beam source (1064 nm wavelength).

parameters such as temperature, poling time, and applied voltage intensity, $\chi^{(2)}$ coefficients of these samples can be shown to be 40–136 pm V^{-1} depending on their structures. High optical nonlinearity of PU1-DCN and PU2-DCN comes from the strong electron-acceptor characteristics of the dicyanovinyl group. The resonance effect of PU1-DCN and PU2-DCN, due to some absorption in their frequency-doubling range, also contributes to the high optical nonlinearity [68]. The low optical nonlinearity exhibited by the PU2-TCN (one more cyano group than PU2-DCN) in comparison to PU1-DCN can be reasoned to be caused by the lower resonance contribution. This lower UV absorption at the frequency-doubling range of 532 nm wavelength as UV absorption shifts to a longer wavelength is known to be the main cause of lower contribution [69].

Stilbazolium salt-type chromophores are some of the most promising candidates for nonlinear optics due to the larger second-order NLO activity and the ease of structure variation [72–75]. They can be applied to nonlinear optics in the form of single crystals [76–79], supramolecular architectures [80,81], or Langmuir-Blodgett (LB) films [82–84]. However, the incorporation of chromophores into the polymer backbone is the most effective means. Notably, the use of a bulky tetraphenylborate counterion in the stilbazolium salt promotes higher optical nonlinearity of the chromophore. Also, it is expected to reduce the ion mobility in a strong electric field during the poling process and thus induce the effective polar alignment of dipoles [65,85]. In fact, PU1-C4B shows large optical nonlinearity of $\chi^{(2)} = 128$ pm V^{-1}. A notable feature of this polymer system is the extended temporal stability of aligned dipoles. At room temperature, no decay in initial SHG activity was exhibited over a period of one month. In addition, 70% of the initial optical activity of the poled sample was maintained at 100 °C, after being exposed for 100 hours. These results imply that the hydrogen bonding between the neighboring polyurethane chains provided a stabilization effect, which in turn prevented the relaxation of oriented molecular dipoles.

In terms of practical applications of organic NLO materials to integrated optics, the fabrication of channel waveguides is important. Some of the well-known techniques for producing channel waveguides include photobleaching, UV lithography, laser ablation, and reactive ion etching. The simplest and most effective means for getting the precise refractive index profile control is known to be the photobleaching method and hence this is utilized for channel waveguide fabrication on PU1-C4B. The refractive index change is measured on exposure to the energy power of 9 kJ cm^{-2} (He-Cd: 442 nm). With photobleached samples, the UV absorption peak height decreases with increasing exposure time. No new peaks can be observed from photobleaching, indicating that unlike azo or stilbene dyes, the mechanism of photobleaching in the stilbazolium salt chromophore is not associated with *cis-trans* isomerization. Twenty and 50 mm-width channel waveguides can be produced with measured propagation loss of the photobleached waveguide of about 1 dB cm^{-1} with a He-Ne laser with the wavelength of 633 nm (Fig. 4a).

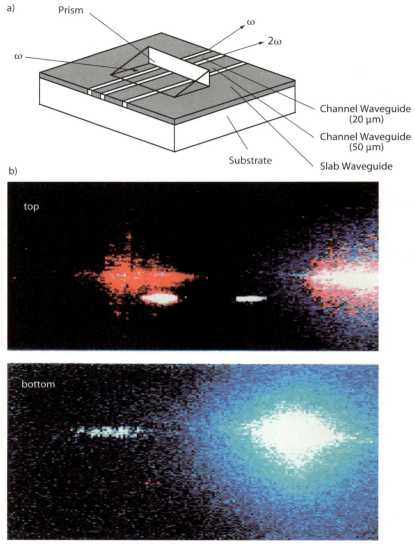

Fig. 4a,b. PU1-C4B waveguides for the Cerenkov-type phase matching SHG (**a**), and phase-matched blue SHG arches (**b**): SHG wave of 380 nm and fundamental wave of 760 nm (*top*) and far field pattern of Cerenkov SHG wave of 400 nm (*bottom*)

Cerenkov-type phase-matched blue SHG in such a produced waveguide is typical (Fig. 4b). The conversion efficiency of this waveguide exhibits at least one-order of magnitude higher than those of the nonphase-matched samples, even where conversion efficiency is not fully optimized. This activity typically lasts for more than

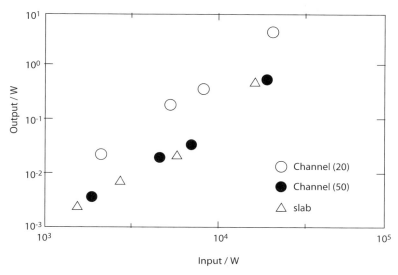

Fig. 5. Conversion efficiency of the phase-matched SHG intensity in PU1-C4B waveguides

two months and the frequency-doubling efficiency of the Cerenkov-type sample is clearly improved relative to the slab-type waveguide (Fig. 5) [86].

3.2.2
Polyimide-Based NLO Polymers

In spite of excellent thermal stability and mechanical properties of wholly aromatic polyimides, the lack of processability has been the major hurdle in using them as the NLO matrix polymers. Polyetherimide (PEI) (Scheme 3), which has an ether linkage in the polyimide backbone, is a better matrix material, as processability and solubility are enhanced without losing the advantageous properties of polyimides. In general, PEI is synthesized by two different processes, that is, nitro-displacement and imidization polymerization [87–91]. However, since this method is not applicable for synthesizing NLO-incorporated PEI, other methods need to be utilized. The alternative synthetic approach for making the NLO polyetherimides is the use of the single-step reaction of NLO-diol and diimide by the Mitsunobu reaction using diethyl azodicarboxylate (DEAD) and triphenylphosphine in anhydrous tetrahydrofuran solvent (Scheme 4) [92]. In this method, the polyimide structure is directly formed during condensation polymerization without a thermal imidization step. When PEI-TH is reacted further with tetracyanoethylene in DMF solvent, a blue-colored copolymer (PEI-TH-*co*-PEI-TCN) is produced, which in some parts contained up to 56% tricyanovinyl group. As the percentage of tricyano (TCN) group increased, the NLO activity also got larger. How-

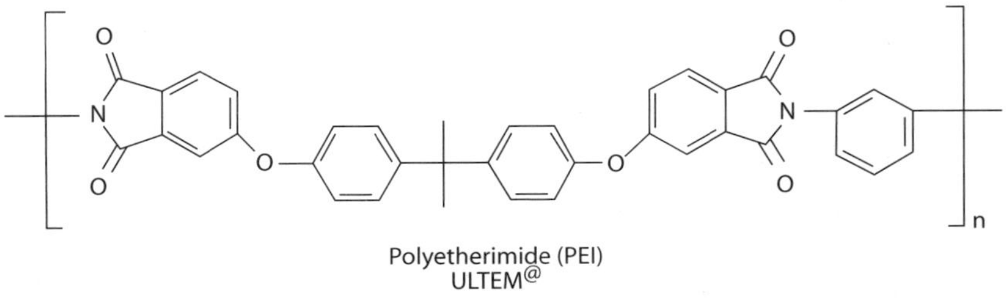

Polyetherimide (PEI)
ULTEM@

Scheme 3

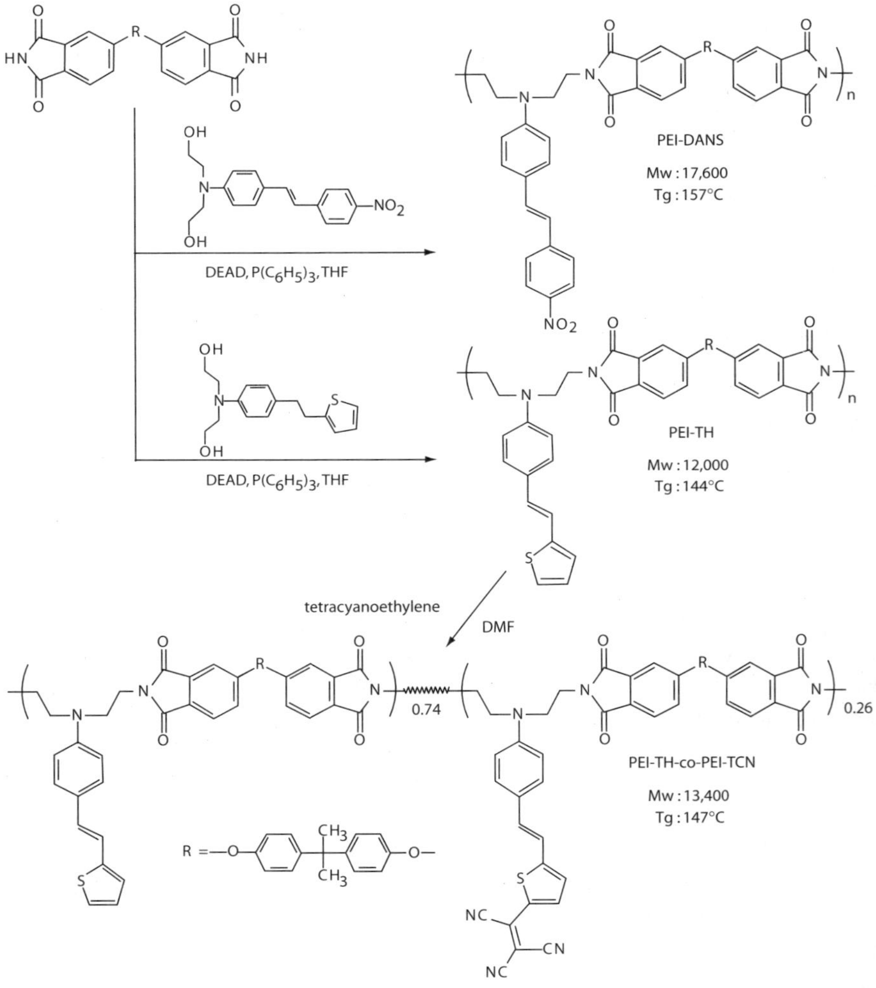

PEI-DANS

Mw : 17,600
Tg : 157°C

PEI-TH

Mw : 12,000
Tg : 144°C

tetracyanoethylene   DMF

PEI-TH-co-PEI-TCN

Mw : 13,400
Tg : 147°C

Scheme 4

ever, solubility decreases with TCN and at concentrations over 40% film, quality became too poor to be applicable.

According to UV-vis spectra, the absorption maximum for the π–π* transition of the stilbene chromophore in PEI-DANS is 441 nm. A new absorption maximum at 669 nm was found for PEI-TH-co-PEI-TCN (not observed in PEI-TH), suggesting that the NLO chromophore is generated by tricyanovinylation. Also, the absorption maximum at 365 nm indicates that the substitution reaction is not complete and some portions remain unreacted with the polymer chain. DSC analysis showed T_g values in the range 144–157 °C for all three PEIs. Not so high T_g values are attributed to the flexible ether linkage and bulky chromophore incorporation to the polyimide backbone. The NLO activity of polymer films on indium tin oxide (ITO) glass was produced by using corona discharge-induced electric poling. The macroscopic $\chi^{(2)}$ values were determined to be 73 pm V^{-1} for PEI-DANS and 56 pm V^{-1} for PEI-TH-co-PEI-TCN. According to UV-vis absorption spectra, the resonant enhancement effect of $\chi^{(2)}$ values may not be significant in both samples, particularly a copolymer in which a very low absorption occurred at 532 nm. The NLO activity of the copolymer system is not high compared to PEI-DANS. However, considering the low degree of substitution of TCN, the $\chi^{(2)}$ of this copolymer is significant. In recent work, some basic knowledge for increasing the substitution

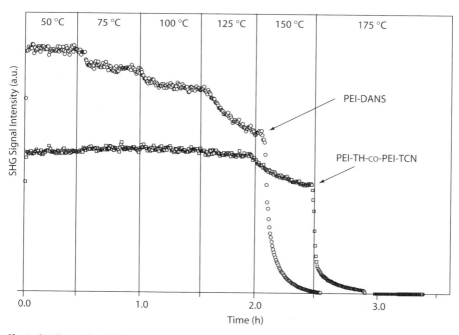

Fig. 6a,b. Thermal stability of aligned dipoles in PEI-DANS (**a**) and PEI-TH-co-PEI-TCN (**b**)

value up to about 56% by controlling polymer reactions has become available [93]. The thermal stability of SHG activities for poled films of PEI-DANS and PEI-TH-*co*-PEI-TCN are shown in Fig. 6. The polymers exhibited quite different thermal stabilities in terms of the poled chromophore relaxation. The SHG signal intensity of PEI-DANS slowly decreased with increasing temperature. This decay process was similar to those of other side-chain NLO polymers reported [94]. However, the SHG signal intensity for the copolymer system did not change (within experimental measurement) from its initial value up to 125 °C. Moreover, the initial NLO activity was sustained even after four weeks at 100 °C. The much-enhanced thermal endurance of aligned dipoles is comparable to that of crosslinked NLO polymers. This unusual thermal stability shown by the copolymer may be attributed to an attractive interaction between nonbonding electrons of thiophene and strongly electron-withdrawing TCN group, through which the improved thermal stability is provided.

In the case of PEI-based NLO polymers, T_g is not very high due to the ether linkage. Therefore, to obtain thermally stable NLO polymers through the enhancement of T_g, utilization of a rigid chain polymer like NLO polyimide would be useful. However, wholly aromatic PI structures can reduce processability as already mentioned. Hence, to alleviate such problems a short aliphatic spacer and hexafluoroisopropyl unit for disrupting linearity was used in NLO polyimide synthesis [95–97]. To introduce the NLO-active chromophore on polyimide matrix, AZO-OH$_2$ and 6F-DI were first prepared by the classical synthetic method. After the polymerization of these two components by using the Mitsunobu reaction, the dioxolane group of the resulting polymer PI-PRO was deprotected to give the polymer PI-DEP with a free aldehyde group, and this polymer was further reacted

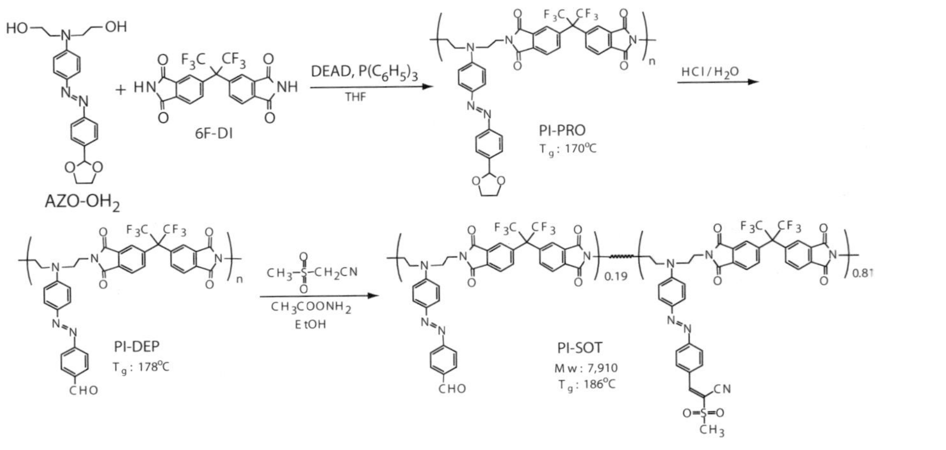

Scheme 5

with methansulfonylacetonitrile to produce the final polymer PI-SOT (Scheme 5) [96]. From the DSC thermogram, the T_g of this copolymer with the substitution degree of 81% cyanosulfonyl chromophore was observed at 186 °C. The absorption maximum was exhibited at 421 nm. Under the conditions of 5 kV poling voltage applied to the corona needle at 186 °C for 10 min, the order parameter value ($f = 1 - A_1/A_0$; A_0 and A_1 are absorbances of the PI-SOT film before and after corona poling, respectively) for PI-SOT was estimated to be 0.22.

From the Maker fringe technique, we obtained a d_{31} value of 50 pm V^{-1} (contains resonant effect due to some absorption in SHG range) for PI-SOT. Since the relationship $d_{33} = 3d_{31}$ holds in the polymer systems, the d_{33} value is supposed to be about 150 pm V^{-1}. From the calculation with the approximate two-level model, the nonresonant value ($d_{33}(\infty)$) of this copolymer was found to be about 47 pm V^{-1}. Figure 7 shows the temperature-dependent SHG intensity for the purpose of the thermal stability study. The SHG signal is quite stable until the temperature reaches 150 °C. The E-O coefficient (r_{33}) was also measured by an ellipsometric reflection technique. The r_{33} value observed for the PI-SOT film was 28 and 5.4 pm V^{-1} for the incident laser 633 and 852 nm wavelengths, respectively. At room temperature in air, the temporal stability of the E-O coefficient of the film was maintained over 95% of its initial value, even after 60 days. Moreover, at 100 °C an initial decay of about 10% was observed within one hour, but no further decay in E-O coefficient was shown within the prolonged time of 30 days. This study confirmed the role of polyimide as a high-temperature NLO polymer matrix, as thermal stability is greatly enhanced.

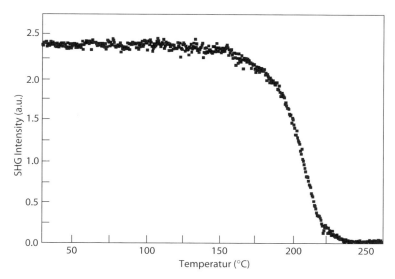

Fig. 7. Thermal stability of aligned dipoles in PI-SOT

3.3
Crosslinked NLO Polymers

To prevent the relaxation of aligned dipoles in NLO polymers, the most effective method is to align NLO chromophores first and then crosslink either the polymer backbone or the chromophores together. Many investigators have attempted such a method and it has shown to greatly increase the long-term stability of the NLO activity [98–102]. However, in most crosslinked polymer systems the optical losses caused by the limited uniformity of the crosslinked reaction are significant. To circumvent these problems, novel crosslinking reaction techniques proposed by Park et al. [103] and Lee et al. [104] have been investigated. In these methods a liquid polymer system with high flow property, poly[(phenyl isocyanate)-co-formaldehyde] (PPIF), which has one crosslinking site located at every phenylene unit, was used for improving the uniformity of the crosslinking reaction between the polymer chains and hydroxy-functionalized NLO chromophores. The synthetic routes

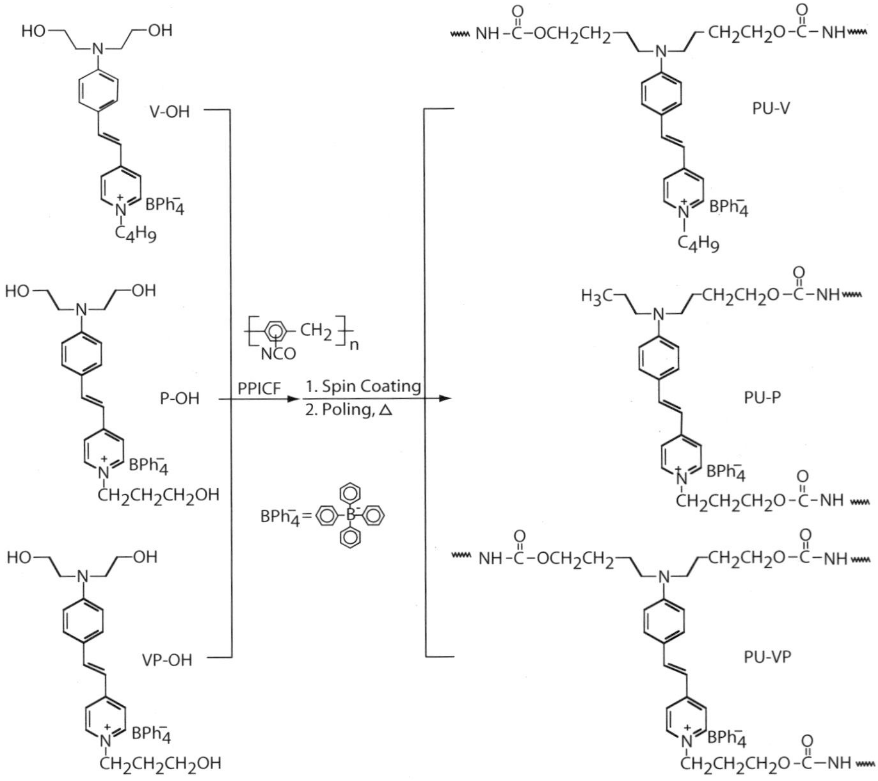

Scheme 6

to the crosslinked polymers with three different bonding sites are shown in Scheme 6. The dried monomers were dissolved in the mixed solvent of DMF and cyclopentanone and the diluted PPICF was added and mixed thoroughly for 10 min. The resulting mixture was then spin-coated. For PU-VP, the solutions did not exhibit any flow characteristics due to fast gelation. The films were directly poled and their SHG signal was measured at the in situ poling set-up. Here, thermally induced condensation reactions of hydroxy functionalities and isocyanate groups lead to a hardened lattice during subsequent induction of polar alignment by the electric field. The T_g of these crosslinked polymers did not appear below the decomposition temperature of chromophores. According to SHG measurement data, relatively high second-order optical activity was shown for the poled and

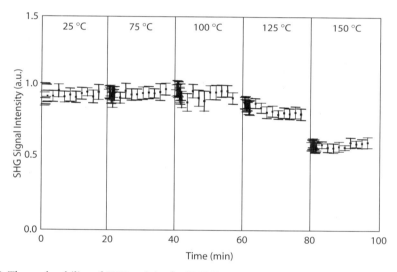

Fig. 8. Thermal stability of SHG activity for PU-VP

Table 3. Thermal stability data of $\chi^{(2)}$ coefficients for crosslinked polyurethanes with different bonding directions

Samples[a]	Bonding direction[b]	$\chi(2)$ (pm V^{-1})			
		25 °C	75 °C	100 °C	125 °C
PU-V	Vertical	37.4	31.8 (85%)[c]	20.6 (55%)	4.9 (13%)
PU-P	Parallel	16.8	15.0 (90%)	9.6 (57%)	3.5 (21%)
PU-VP	Vertical/Parallel	30.2	30.2 (100%)	30.2 (100%)	26.9 (89%)

[a]Sample films were cured at 100 °C for 12 h under vacuum. [b]Bonding direction indicates the approximate angle between dipole of chromophore and two bonding sites. [c]Values in parenthesis indicate retained percentage of initial NLO activity at the corresponding temperature.

cured polymer films. The $\chi^{(2)}$ values for PU-V, PU-P, and PU-VP were 37.4, 16.8, and 30.2 pm V^{-1}, respectively. The lower NLO activity of PU-P is indicative of a significant reduction of parallel locking structure due to crosslinking effects [105]. Figure 8 represents the thermal stability of SHG activity for PU-VP evaluated during the stepwise temperature increase from room temperature to 150 °C. Results for all samples are also listed in Table 3. The decay of oriented dipoles in PU-VP was minimized due to lattice hardening of PU with three bonding sites, as expected. Also, the SHG intensity does not show any reduction until the temperature rises to 100 °C. For temperatures below 100 °C, the $\chi^{(2)}$ values for PU-V and PU-P decreased to 85–90% of the initial value [104, 106].

3.4
Hyperbranched and Dendritic Polymers

Dendrimers are a relatively new class of macromolecules different from the conventional linear, crosslinked, or branched polymers. Dendrimers are particularly interesting because of their nanoscopic dimensions and their regular, well-defined, and highly branched three-dimensional architecture. In contrast to polymers, these new types of macromolecules can be viewed as an ordered ensemble of monomeric building blocks. Their tree-like, monodispersed structures lead to a number of interesting characteristics and features: globular, void-containing shapes, and unusual physical properties [107–111].

3.4.1
Spontaneous, Noncentrosymmetric Organization of NLO Chromophores in Dendritic Structures (NLO Dendritic Effect at the Molecular Level)

To study the dendritic effect of dendrimers on NLO properties, a series of dendritic macromolecules, such as the azobenzene-containing dendrons **9** (Scheme 7), have been developed and subjected to analysis of their conformational and molecular NLO properties [112, 113]. These molecules were modified by introducing 1–15 numbers of azobenzene branching units as the NLO chromophore and by connecting these units with aliphatic chains at the end of dendritic structures. In these topologically complex molecules, each chromophore contributed coherently to the macroscopic NLO activity. The molecular hyperpolarizability (β) of the azobenzene dendron with 15 chromophoric units was measured to be 3,010 × 10^{-30} esu by the hyper-Rayleigh scattering method. This value is approximately 20 times greater than that of the β for the individual azobenzene monomer (150 × 10^{-30} esu). In addition, the structural information on dendrons provided by the polarized NLO measurement also indicated that each chromophore was oriented noncentrosymmetrically along the molecular axis to become a cone shape rather than a spreading or a spherical shape. This structure gave rise to a large elec-

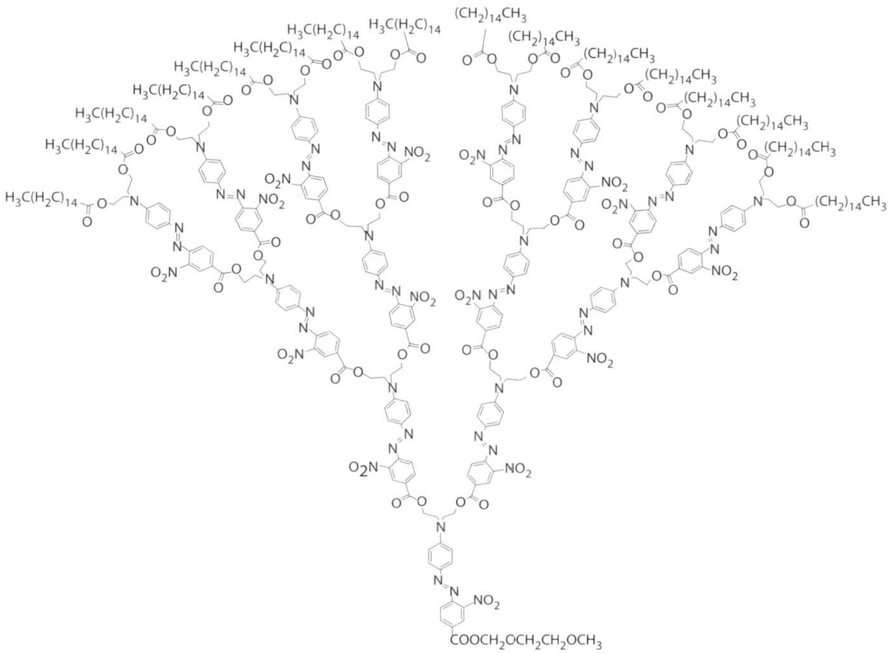

Scheme 7

tronically dipolar macromolecular system, in which each chromophoric unit coherently contributed to the second harmonic generation.

By extending the above observations on dipolar dendrimers, Ledoux and Zyss et al. [114] have recently reported the first evidence of spontaneous supramolecular organization of octupolar molecular units, resulting in a quasi-optimized, acentrically ordered dendritic structure. Having evaluated β for the monomer, size effects can be investigated for the different polymetallic species **10–12** (Scheme 8), in view of the similarity of their absorption features. A large increase of β was clearly observed for the $N = 7$ dendrimer relative to that of the monomer. The corresponding β value was significantly higher than that measured for $N = 14$ linear polymers. It is interesting to plot β as a function of the number of monomers (N), as displayed in Fig. 9a. Comparison between the heptamer ($N = 7$) and polymer ($N = 14$) values highlights strikingly different behaviors. The β values for $N = 14$, 3, and 2 fit almost perfectly a β (polymer) = $N^{1/2}\beta$ (monomer) scaling law, whereas a dendritic heptamer ($N = 7$) satisfactorily fits a linear relationship β (dendrimer) = $N\beta$ (monomer). Such a quasi-linear dependence is the clear signature of a quasi-optimized octupolar ordering of individual building blocks in a dendrimer, in

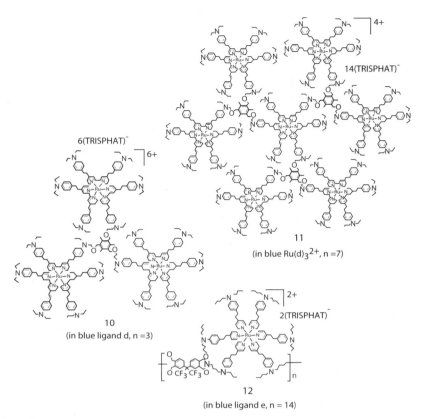

Scheme 8

contrast with the fully disordered structure expected for a linear polymer, as inferred from the $N^{1/2}$ dependence of the corresponding β value. The NLO elementary units in the linear polymer do not display any acentric organization, and the corresponding harmonic light scattering (HLS) response of the polymer is similar to that of a set of 14 independent molecules in a solution. Conversely, a semi-rigid, optimized acentric organization results in coherent second harmonic emission at the supramolecular level; the corresponding intensity from the dendrimer solution is almost proportional to $N^2_{monomer} < \beta^2_{monomer} >$. The $N^2_{monomer}$ dependence assumes a linear additive model, whereby the individual sub-units follow an interaction-free oriented gas behavior.

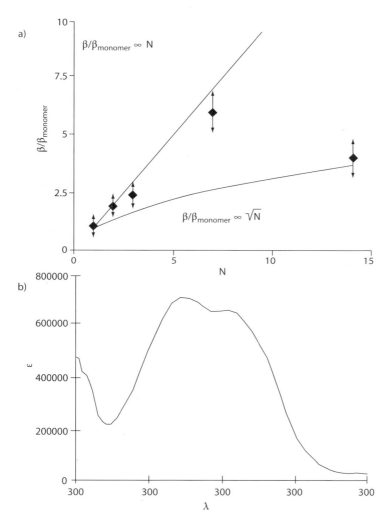

Fig. 9a,b. Plot of the ratio $\beta/\beta_{monomer}$ as a function of the number of ruthenium complex building blocks N. ◆: experimental data; *continuous lines* display the N or $N^{1/2}$ dependence of the ratio of the oligomer ($N = 2, 3, 7,$ and 14) hyperpolarizability β over that of the monomer $\beta_{monomer}$ (i.e., $\beta/\beta_{monomer}$). *Vertical arrows* show the experimental error (**a**), UV-vis spectrum of heptamer **11** (**b**)

3.4.2
Enhancement of Poling Efficiency in NLO Dendrimers (NLO Dendritic Effect at the Macroscopic Level)

In the fabrication of practical E-O devices, all of the three critical materials issues (large E-O coefficients, high stability, and low optical loss) need to be simultaneously optimized. One of the major problems encountered in optimizing polymeric E-O materials is to efficiently translate the large β values of organic chromophores into large macroscopic electro-optic activity (r_{33}). According to an ideal-gas model, macroscopic optical nonlinearity should scale as $\mu\beta/M_w$ (M_w is the chromophore molecular weight) [115]. Therefore, E-O coefficients of many hundreds of picometers per volt should be expected if such molecular optical nonlinearity shown in Table 1 could be effectively translated into macroscopic E-O activity. However, the electric field poling of a polymer containing highly nonlinear chromophores is often hindered by the large dipole moment of the molecules. Theoretical analysis suggests that the maximum realizable E-O activity can be enhanced by modifying chromophores with bulky substituent groups [116]. These groups do not influence molecular hyperpolarizability but they will minimize the effect of unwanted electrostatic interactions by inhibiting the side-by-side approach of chromophores along the minor axes of the prolate ellipsoidal unmodified chromophores. Theoretical analysis also suggests that the ideal chromophore shape is that of a sphere [117].

Globular-shaped and void-containing dendrimer synthesis is well suited to obtain spherical chromophore shapes and realize large E-O coefficients. Dendritic structures can be also used to control solubility and processability, to improve chemical and photochemical stability, and to facilitate lattice-hardening schemes carried out after the electric-field poling process. In addition, by exploiting selective halogenation and isotopic substitution, the large void-containing dendrimers can help minimize optical loss (due to light scattering or vibrational absorption). The recent development of single-dendron-modified NLO chromophores, multiple-dendron-modified NLO chromophores, and multiple-chromophore-containing crosslinkable NLO dendrimers will be highlighted.

3.4.2.1
Single-Dendron-Modified NLO Chromophores

Three kinds of NLO chromophores **13–15** (Scheme 9) were obtained through the modification of one highly nonlinear tricyanofuran (FTC) acceptor-based chromophore with different shapes and sizes of substituents. When compared to the *t*-butyldimethylsilyl side group substituted in chromophore **13**, the dendritic substituent on chromophore **15** has a much more flexible and wide spreading shape because of the branching alkyl chains that are connected to the phenyl

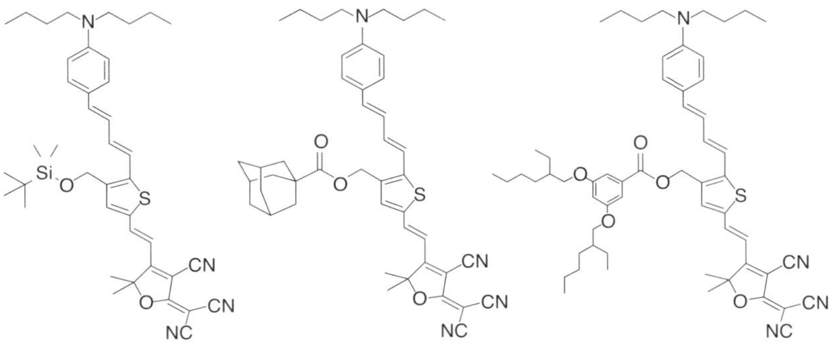

Scheme 9

group. The adamantyl group on chromophore **14** has a more rigid and bulky structure. Due to the existence of an insulating methylene spacer between the active D-π-A moiety and the side group, all these modified chromophores gave essentially the same absorption peak ($\lambda_{max} \approx 641$ nm) in dioxane solution, indicating that these substituents did not affect the molecular hyperpolarizability. However, it is expected that these side groups will affect the poling efficiency and the macroscopic susceptibility (r_{33}) of polymeric materials that are incorporated with these chromophores. Two aspects of the substituent effects may contribute to this: one is that the different shape and size of substituents will lead to substantially different intermolecular electrostatic interactions among chromophores; the other is that the different rigidity and size of substituents will also create variability of free volume which in turn will affect the mobility of the chromophore under high electric field poling conditions.

To study the substituent effect on poling efficiency, all three chromophores were incorporated into an amorphous polycarbonate (APC, $T_g = 205$ °C) matrix with a predetermined loading density and the resulting polymeric materials were processed, poled, and measured with the same condition, respectively. Figure 10 shows the dependence of the r_{33} value on chromophore number density measured for each system. For the chromophore **13**, it exhibited a maximum value of r_{33} and the value decreased with the increase of the chromophore loading level. In addition, it stayed at a lower r_{33} level than the other two systems, suggesting a significantly higher chromophore-chromophore electrostatic interaction than expected. In comparison, chromophore **14** and **15** enhanced the achievable value to a much higher level ($\approx 40\%$ of enhancement) due to the minimization of electrostatic interaction among chromophores. However, chromophore systems **14** and **15** contributed differently to the enhancement of r_{33} (Fig. 10) and poling efficiency (Fig. 11) with respect to similar chromophore number density. Compared to the

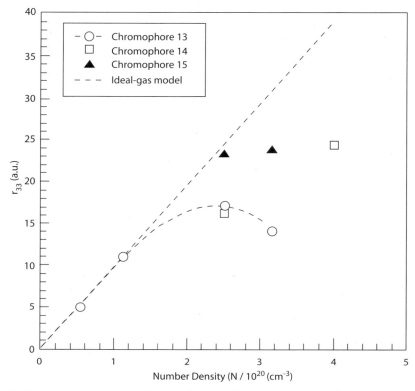

Fig. 10. Comparison on E-O coefficient of NLO polymers incorporated with chromophores containing different loading density and substituent modification

t-butyldimethylsilyl group on chromophore **13**, the dendritic substituent on chromophore **15** offers notable enhancement for the obtainable r_{33} and poling efficiency at lower chromophore number density, whereas the adamantyl substituent on chromophore **14** provides more significant improvement for r_{33} and poling efficiency only at higher chromophore number density. This is probably due to the fact that the flexible and bulky dendritic moiety on chromophore **15** provides more free volume for the chromophore orientation mobility under high electric field poling compared to the rigid adamantyl group on the chromophore **14**.

3.4.2.2
Multiple-Dendron-Modified NLO Chromophores

To maximize the dendritic effect, a chromophore **16** was recently modified further with multiple fluorinated dendrons (Scheme 10). The resulting dendritic chromophore **17** exhibits appreciably different properties, such as a blue-shifted absorp-

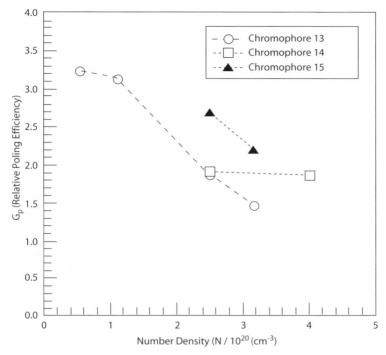

Fig. 11. Comparison on relative poling efficiency of E-O polymers incorporated with chromophores containing different loading density and substituent modification

tion (29 nm), a higher thermal stability (20 °C), and a much higher E-O activity (×3) (Fig. 12), compared to the pristine chromophore **16** in the same polymer matrix (Table 4). Since both chromophores contain the same D-π-A active moiety, the blue-shifting of the absorption λ_{max} is probably due to the hydrophobic and lower dielectric environment created by the perfluorodendrons. Encapsulating the NLO chromophore with several fluorinated dendrons also improves its thermal decomposition temperature. In the context of studying dendritic effect on macroscopic E-O activity, both chromophores were incorporated into the APC polymer with a predetermined loading density (12% wt.) and the resulting guest-host materials were processed, poled, and measured under the same condition, respectively. Amazingly, a three times larger E-O coefficient in the **17**-doped system was observed. This proves that the lower dielectric environment and the site-isolation effect provided by the dendrons greatly facilitate the poling process.

Currently, several more efficient NLO chromophores such as the chromophore **18** are being modified with the similarly fluorinated dendrons to provide further improved macroscopic E-O properties.

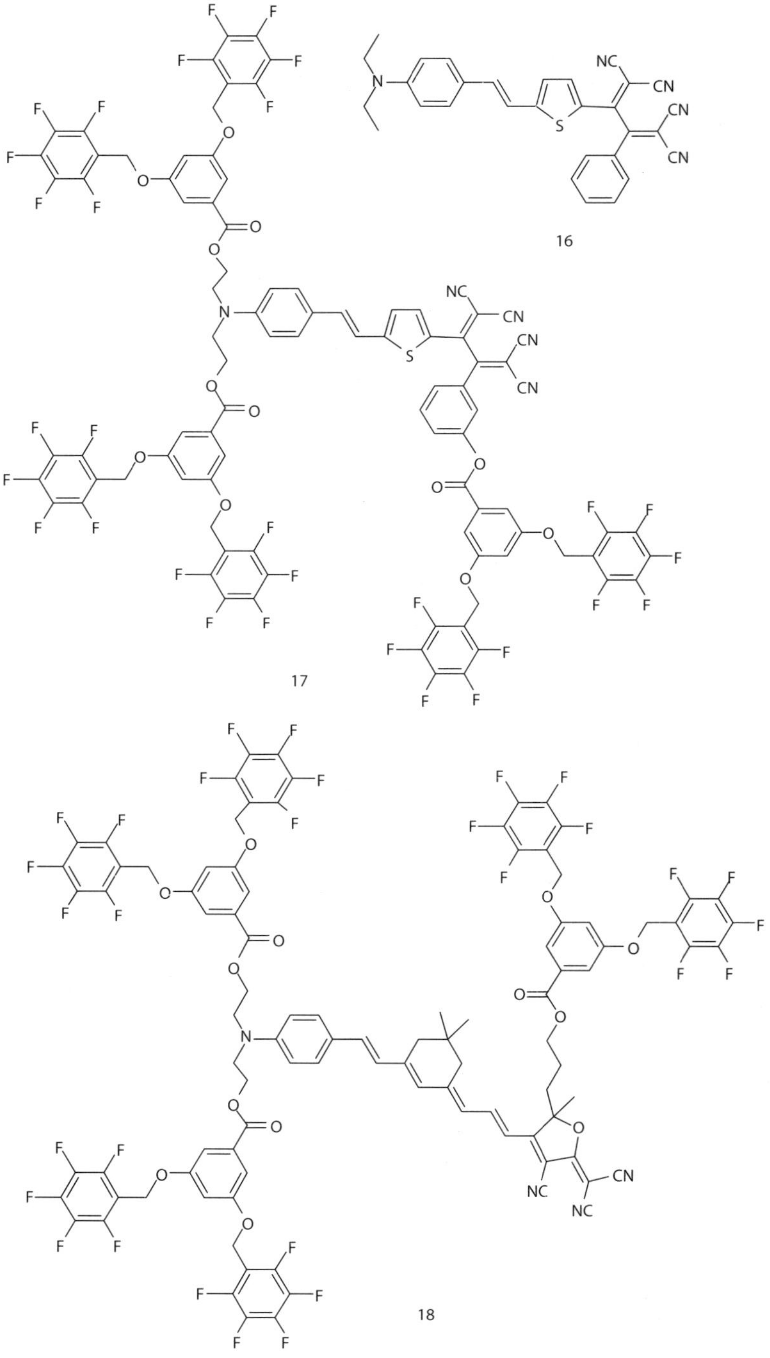

Scheme 10

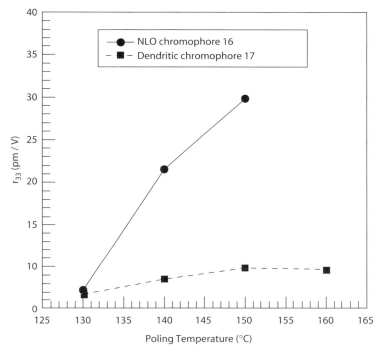

Fig. 12. Comparison on E-O coefficient of NLO polymers incorporated with non-dendritic chromophore **16** and dendritic chromophore **17** at different poling temperatures (12% wt. of chromophore in APC, 1.0 MV cm^{-1} of poling field)

Table 4. Comparison of the properties of NLO non-dendritic chromophore 16 and dendritic chromophore 17

	λ_{max} (1,4-dioxane) (nm)	T_d[a] (°C)	r_{33} at 1.3 μm[b] (pm V^{-1})
NLO chromophore **16**	603	245	10
Dendritic chromophore **17**	574	265	30

[a]Determined by DSC at the heating rate of 10 °C min^{-1} under nitrogen. [b]Poled at 150 °C under the field of 1.0 MV cm^{-1} with 12% wt. of NLO chromophore in APC.

3.4.2.3
Multiple-Chromophore-Containing Crosslinkable NLO Dendrimers

To create an ideal stand-alone and spherical shape NLO material that does not involve the complexity of incorporating NLO dye in a separate polymeric matrix, multiple NLO chromophore building blocks can be further placed into a dendrim-

o   : Core moiety

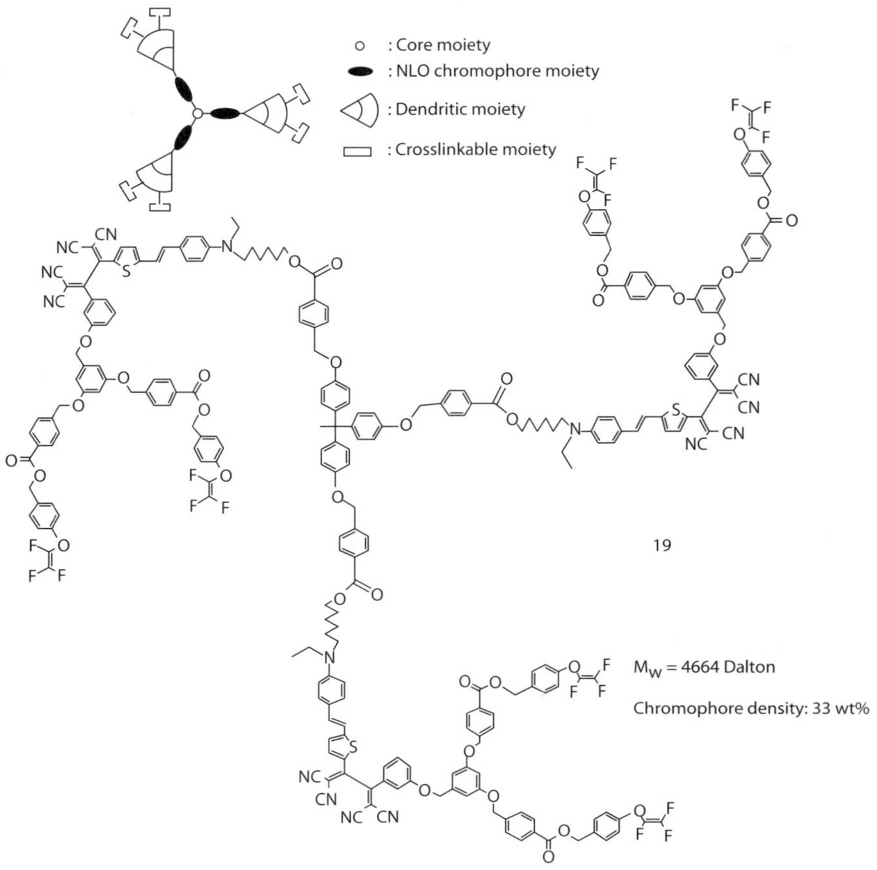

●   : NLO chromophore moiety

◁| : Dendritic moiety

☐   : Crosslinkable moiety

19

M_W = 4664 Dalton

Chromophore density: 33 wt%

Scheme 11

er to construct a precise molecular architecture with a predetermined chemical composition. A crosslinkable NLO dendrimer **19** (Scheme 11) which has been developed for this purpose exhibits a very large optical nonlinearity and excellent thermal stability [118]. The NLO dendrimer was constructed through a double-end functionalization of the three-dimensional Ph-TCBD thiophene-stilbene-based NLO chromophore **16** as the center core and the crosslinkable trifluorovinylether-containing dendrons as the exterior moieties. Spatial isolation from the dendrimer shell decreases chromophore-chromophore electrostatic interactions, and thus enhances macroscopic optical nonlinearity. In addition, the NLO dendrimer can be directly spin-coated without the usual prepolymerization process needed to build up viscosity, since it already possesses a fairly high molecular weight (4664 Da). The dendrimer also possessed a quite high chromophore loading density (33% w/w) and showed no indication of any phase separation due to the incompatibility between the chromophore and the peripheral dendrons. There are also several other advantages derived from this approach, such as excellent alignment stability and mechanical properties, which are obtained through the sequential hardening/crosslinking reactions during the high-temperature electric-field poling process. A very large E-O coefficient ($r_{33} = 60$ pm V^{-1} at 1.55 μm) and long-term alignment stability (retaining >90% of its original r_{33} value at 85 °C for more than 1000 h) were achieved for the poled dendrimer. In comparison, E-O studies were performed on the guest/host system in which nondendron-modified, similarly structured chromophore **16** (optimized loading level: 30% wt.) was formulated into a high-temperature polyquinoline (PQ-100). The T_g of the resulting system is plasticized to approximately 165 °C. After the same sequential heating and poling as that of the dendrimer, the guest-host system showed a much smaller E-O coefficient (less than 30 pm V^{-1}) and worse temporal stability (only retained <65% of its original r_{33} value at 85 °C after 1000 h) (Fig. 13). In addition, the attempt to corona pole the nontrifluorovinyl ether functionalized dendrimer only

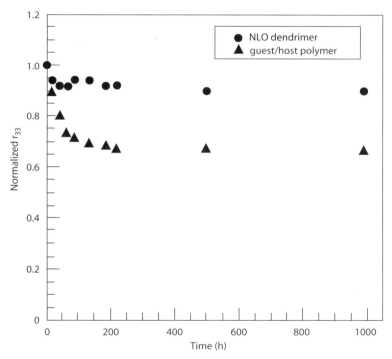

Fig. 13. Temporal stability of the poled/crosslinked NLO dendrimer **19** and guest/host polymer system (NLO chromophore **16**/PQ-100) at 85 °C in nitrogen. Normalized r_{33} as a function of baking time

showed a very fast decay of E-O signal ($<$10 pm V^{-1}) after the same sample being poled and measured at room temperature. This is due to the intrinsically low T_g ($<$50 °C) and very large free volume of the dendrimer. On the basis of these results, the large r_{33} of the poled dendrimer is largely due to the dendritic effect that allows the NLO dendrimer to be efficiently aligned. On the other hand, the high temporal stability of the poled dendrimer mainly results from the efficient sequential crosslinking/poling process.

To conduct a more comprehensive study on E-O dendrimers, a series of crosslinkable NLO dendrimers with different cores, branches (with more active chromophore components), shape, and size has been developed. For example, the dendrimer **20** (Scheme 12) was designed and synthesized to further minimize the chromophore-chromophore electrostatic interaction with more dendrons. The shape and the size of the dendrimer was also designed not only to optimize dendron folding under a high electric field but also to improve the processability through higher molecular weight and better compatibility among the core, branches, and periphery of the resulting dendrimers. The preliminary data showed that the increased numbers of dendron in the NLO dendrimers will red-shift the absorption λ_{max} both in its solution and solid forms due to a more polar environment (carboxylate-containing dendrons) around the chromophores (Table 5). However, the encapsulation of NLO chromophores with more dendrons

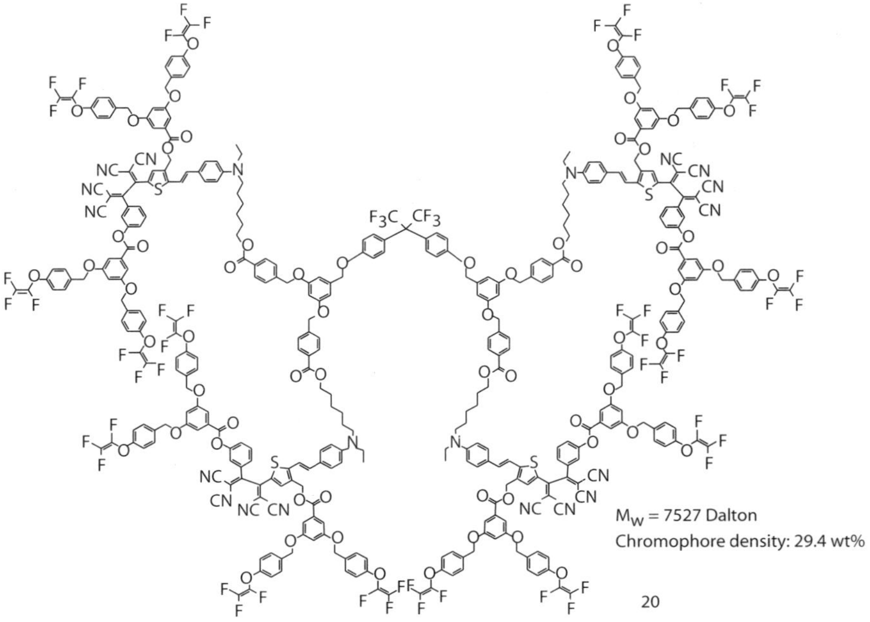

M_W = 7527 Dalton
Chromophore density: 29.4 wt%

20

Scheme 12

Table 5. Comparison on the linear optical properties of NLO chromophore 16 and dendrimers 19 and 20

	λ_{max} (nm) In solution (1,4-dioxane)	In film
NLO chromophore **16**	603	646
NLO dendrimer **19**	608	655
NLO dendrimer **20**	625	662

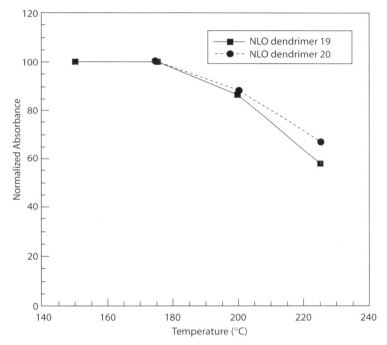

Fig. 14. Comparison on thermal stability of NLO dendrimers **19** and **20** with the isotherm heating/150, 175, 200, and 225 °C for 20 min under nitrogen atmosphere, respectively. Normalized absorbance as a function of baking temperatures

also protects active sites on the NLO chromophore from reacting with each others on heating, thus providing better thermal stability (Fig. 14).

A much longer and more efficient tricyanofuran-based NLO chromophore has also been introduced into the NLO dendrimers **21** (Scheme 13), **22**, and **23** (Scheme 14) in which the crosslinkable groups were functionalized at either the acceptor side or at the donor-end positions. It provides great potential not only to further enhance the macroscopic E-O activity but also to achieve the combination of desirable properties.

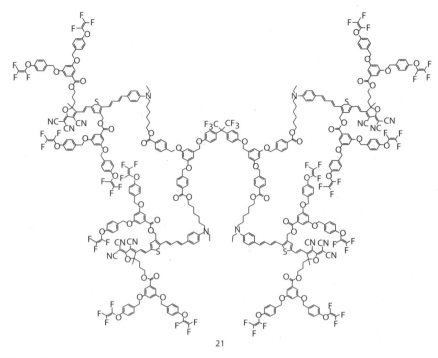

21

Scheme 13

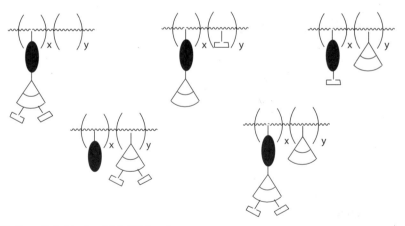

Fig. 15. Crosslinkable dendritic NLO polymers

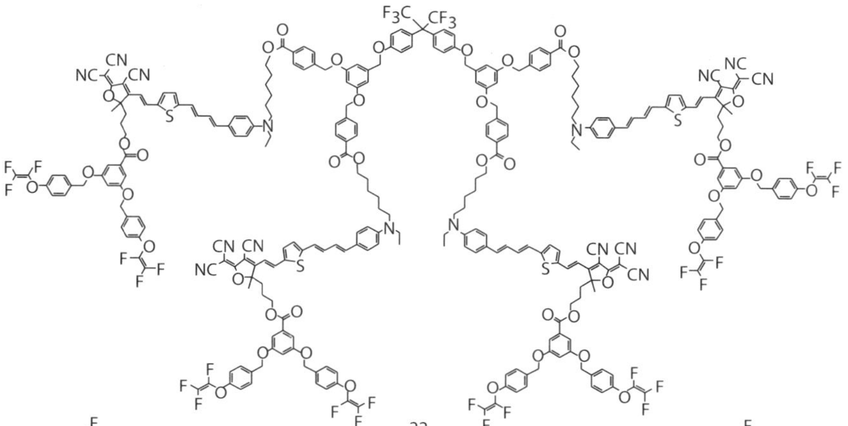

Scheme 14

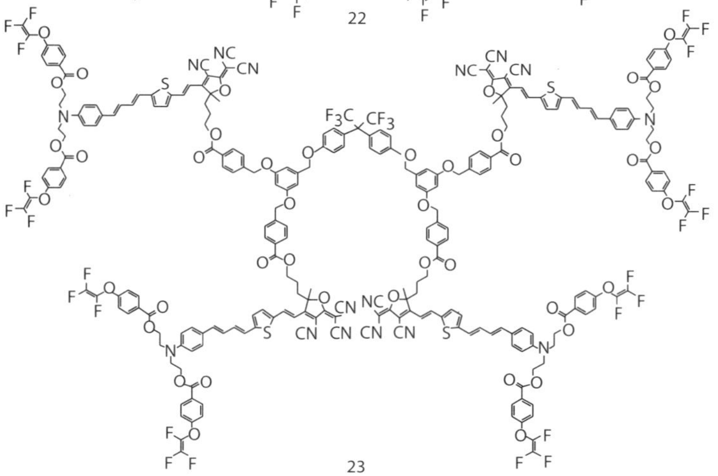

In conclusion, compared to common NLO polymers, dendritic NLO materials provide great opportunities for the simultaneous optimization of macroscopic electro-optic activity, thermal stability, and optical loss. For the dendron-modified NLO chromophore/polymer systems, the interaction between chromophore and high-T_g aromatic polymer should be noteworthy in addition to the chromophore-chromophore electrostatic interactions. For the crosslinkable NLO dendrimers, the nanoenvironment such as dendron shape, size, dielectrics, rigidity and chirality, hydrophilicity/hydrophobility, and the distribution of chromophores in the NLO dendrimers will play very critical roles in achieving the maximum realizable macroscopic properties. Another interesting approach is the employment of the hybrid crosslinkable denendritic NLO polymers. By combining the good processability of linear polymers with the site-isolation effect of dendrimers, the strategy of using hybrid crosslinkable dendritic NLO polymers (Fig. 15) will provide even more flexibility in designing suitable molecular structures to realize high-performance E-O materials.

3.5
Organic-Inorganic Hybrid Materials

Most NLO chromophores are not good photonic materials due to large absorption and high optical losses. Inorganic glasses, however, are excellent photonic media because of high optical quality and extremely low optical losses.

Therefore, hybridization of organic chromophores and inorganic glasses can be one of the best ways to achieve optical materials with large NLO activity and low optical losses. In addition, the use of highly crosslinked silica matrix can reduce thermal relaxation of the aligned dipoles. Taking advantage of these facts, an organic NLO chromophore-inorganic glass hybrid for NLO applications can be made by using sol-gel techniques. In the sol-gel process, the three-dimensional silica network is formed at low temperatures without any thermal decomposition of organic chromophore [119–129]. NLO chromophores are mixed at the molecular level with the silica or anchored to the silica matrix. The resulting system is called organic-inorganic hybrids materials.

The early organic-inorganic hybrid materials reported by Zhang et al. [130] was a dye-doped system (guest-host type) in which the NLO chromophore based on N-(4-nitrophenyl)-(S)-prolinol (NPP) was physically mixed into the sol-gel silica glass. They achieved a moderate $\chi^{(2)}$ value of 10.9 pm V^{-1} and 80% of initial value maintained for more than three months at room temperature. However, the highest mixing ratio without any phase separation of guest molecule into the silica matrix was limited to less than 15% wt. Also, fabrication of sol-gel films thicker than 1 µm was difficult because of severe cracks.

To prevent such problems, a dye-attached sol-gel system in which the NLO chromophore was covalently bonded to the silica matrix by the sol-gel reaction for dye-attached system is illustrated in Scheme 15 [131–133]. By hydrolysis and condensation reactions from the alkoxysilane-functionalized NLO chromophore, dye-attached sol-gel films are obtained. According to the AFM image of all dye-attached sol-gel system, the surface is relatively flat and defects could not be observed. In general the surface roughness of the poled films is larger than the unpoled films. However, in either case the roughness is less than 10 nm and should not be a problem in waveguide applications. It is believed that this remarkable improvement in the quality of the dye-attached film is due to the chemical bonding of the chromophore molecule to the silica matrix, which prevents the dye molecule from aggregating. Also, the chromophore molecules provide flexibility between the stiff silica backbones. E-O coefficients (r_{33}) of the hybrids ranged between 4.7–14.0 pm V^{-1} at 1.3 µm wavelength, as listed in Table 6.

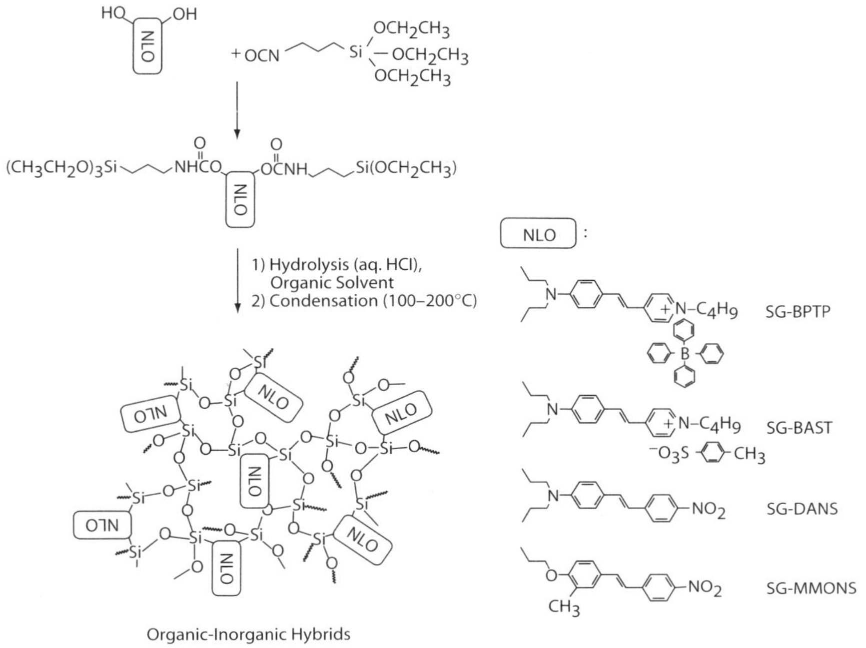

Organic-Inorganic Hybrids

Scheme 15

Table 6. Poling conditions and EO coefficients of various organic-inorganic hybrid films

Samples	λ_{max} (nm)	Poling condition	Applied voltage (kV)	n (TE) at 1.3 μm	r_{33} at 1.3 μm (pm V^{-1})
SG-BPTB	473	160 °C/2 h	13	1.6720	5.0
SG-BAST	465	160 °C/2 h	13	1.6640	4.7
SG-DANS	443	170 °C/3 h	14	1.6102	12.0
SG-MMONS	373	180 °C/2 h	6	1.5974	14.0

By using the hybrid material SG-DANS, a Mach-Zehnder modulator was fabricated [133]. Figure 16 illustrates the top and the cross-sectional views of the E-O modulator. The device was fabricated on an IC-grade silicon substrate on which Cr/Au was deposited as a bottom electrode. For the lower and upper claddings, UV-cured epoxy film was spin-coated to a thickness of 2 μm and the NLO-active layer was made of an NLO dye-silica hybrid film of 2.1 μm thickness. The operating characteristics of the device were examined by using a diode layer of 1.3 μm wavelength. Figure 17a shows the TM single mode pattern in the channel waveguide, which was monitored by CCD camera. Figure 17b presents the inten-

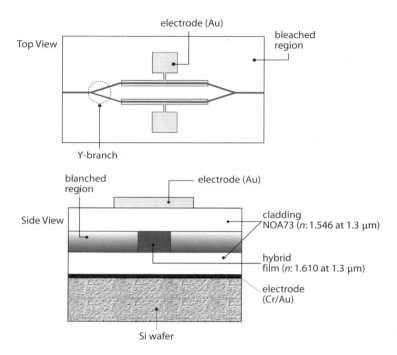

Fig. 16. Schematic diagram of Mach-Zehnder modulator

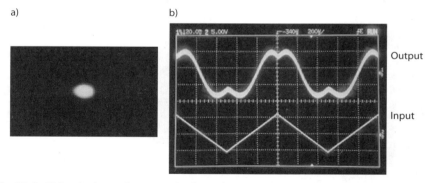

Fig. 17a,b. TM_{00} single-mode pattern in channel waveguide (**a**) and intensity modulation response of Mach-Zehnder modulator (**b**)

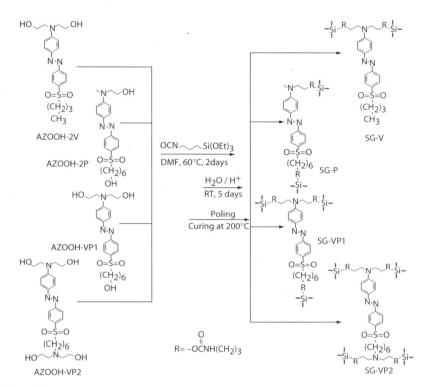

Scheme 16

sity modulation response and applied voltage. The half-wave voltage (V_π) of the modulator was 13 V and the r_{33} value in the waveguide was estimated to be about 10 pm V^{-1}.

To improve the materials' properties including thermal and temporal stability, we have recently developed hybrid materials, which have several fixing sites in the chromophore for the purpose of multiple site attachment to silica matrix. They are characterized as vertical (V), parallel (P), and vertical-parallel (VP) types, depending on the bonding direction between chromophores and silica matrix (Scheme 16) [134]. The measured d_{33} values of the poled hybrid films were 58, 50, 53, and 64 pm V^{-1} at 1064 nm wavelength for SG-V, SG-P, SG-VP1, and SG-VP2, respectively. For the thermal stability investigation of NLO activity, temperature-dependent SHG experiments were carried out in situ by monitoring the SHG signal while heating at a rate of 2 °C min^{-1}. Figure 18 shows the dynamic stability for hybrid films that were poled for 1 hour at 200 °C. It was found that the thermal stability of SG-VP2 with four fixing sites to silica glass matrix was much enhanced. Up to 120 °C practically no decay was observed, indicating that the lattice harden-

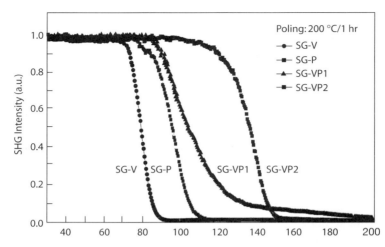

Fig. 18. Thermal stability of SHG activities for organic-inorganic hybrids with different bonding directions

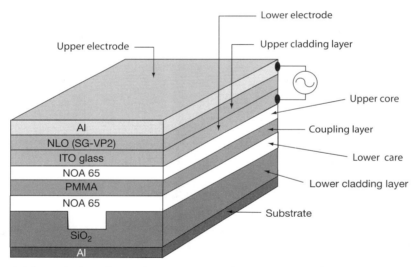

Fig. 19. Schematic view of vertical asymmetric coupled EO modulator fabricated by using SG-VP2

ing of this hybrid with four bonding sites can clearly contribute to improving thermal stability of aligned dipoles. By using this hybrid material, a vertically asymmetric-coupled E-O modulator with antiresonant slab waveguide has been fabricated (Fig. 19). The far field image at the output of the vertical asymmetric coupler and the intensity modulation response by the applied voltage are shown in Fig. 20.

a)

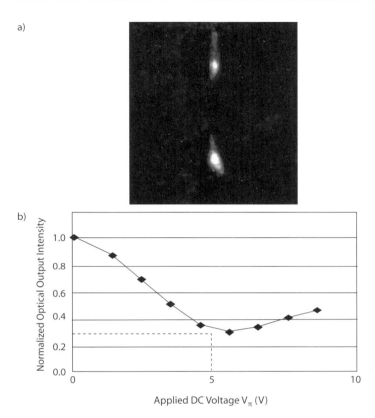

b)

Applied DC Voltage V_π (V)

Fig. 20a,b. Far field mode pattern: the *upper spot with a long stripe line* is the slab-mode pattern and the lower one is the rib of waveguiding propagation mode (**a**), and output intensity modulation as a function of applied DC voltage (**b**)

The resulting device with total device length of 8 mm exhibited a good performance with an extinction ratio of approximately 9 dB, operating voltage of 5 V, optical loss of 0.2 dB cm^{-1}, and insertion loss with 3.5 dB cm^{-1} [135]. Considering that temporal and thermal stabilities of this hybrid system are excellent, the E-O device fabricated by hybrid material has great potential for opto-electronic applications.

4
Chromophore Orientation Techniques

4.1
Static Field Poling

One of the important tasks in getting efficient materials based on functionalized polymers for second-order NLO applications is the creation of macroscopic non-centrosymmetry. This is done by orienting dipole moments in a privileged direction, which is defined by an external electric field. Several techniques have been developed, which use the interaction strength of dipole moment with an external electric field to orient them. These are:
(i)   static field poling
(ii) photoassisted poling [136]
(iii) all optical poling [137].

While the first technique takes advantage of the interaction between dipole moments and the applied static field, the two others take account of the cooperative effect of static and optical field (photoassisted poling) and purely optical field (all optical poling) and photoisomerization process. Between the DC poling techniques we note:
(i)   contact (electrode) poling
(ii) corona poling [138, 139]
(iii) photothermal poling [140, 141]
(iv) electron beam poling [142, 143].

The techniques have some advantages and some drawbacks [144]. The corona poling technique allows application of high poling fields, but often leads to the dielectric damage to thin film and more particularly to its surface, resulting in an increase of the propagation losses after poling.

In the contact poling a large poling field is created through electrodes with the polymer thin film placed in between (Fig. 21). The film is heated up to the glass

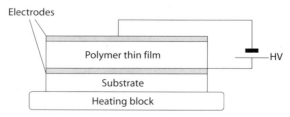

Fig. 21. Schematic representation of an electric contact poling set-up

transition temperature and the poling voltage is applied through electrodes. The main drawback of this technique is the limited voltage that can be applied due to the dielectric breakdown in surrounding air or in the poled film itself [145, 146]. For this reason the poling is usually done in a high-vacuum chamber.

In the corona poling technique, the high poling field in thin film is created by charges deposited on the film surface during the gas ionization in the surrounding discharge needle atmosphere. The film is again heated to the glass transition temperature with a heating block. Sometimes a metallic grid (Fig. 22) is used to control the poling current (Fig. 23) and to get a better homogeneity of poling. While in the electrode poling technique the poling occurs at any applied voltage (although its efficiency depends on it), the corona poling is a threshold phenomenon

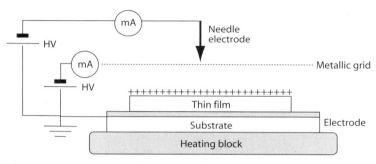

Fig. 22. Schematic representation of a corona poling set-up

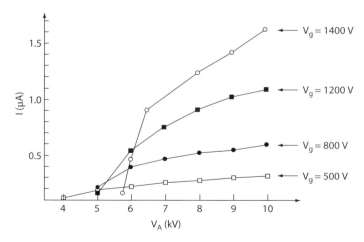

Fig. 23. Variation of the poling current I with the needle voltage V_A for different grid V_g tensions in corona poling technique [151]

as seen in Fig. 23 and usually requires application of relatively high voltages (around 6–8 kV) with the needle electrode at 2–2.5 cm from the thin film surface.

Photothermal poling [140, 141] is a simple modification of the electrode poling technique. The only difference is the use of a laser beam, with wavelength lying in the material absorption band, to heat the thin film. The main advantage of this technique consists of a very localized poling. It has been used for the fabrication of bidirectionally poled polymer films [141].

In the electron beam poling technique [142, 143] a constant current, created in material by a monoenergetic electron beam with an energy of 2–40 keV, is used to orient chromophores. The dipoles are oriented by the polarization field, created by being trapped in bulk-decelerated electrons. In this sense, the technique is very similar to the corona poling technique, with the difference that in the former the poling is due to the field created by the surface, while in the latter it is due to the bulk charges. The technique also allows poling of very small areas and has been used for the fabrication of periodical structures for quasi-phase-matched (QPM) second harmonic generation [147] applications. In guest-host polymer systems PMMA-DR 1, similar poling efficiency to the corona poling technique has been reported [148].

In all cases the use of an electrode is required, with all possible negative aspects such as charge injection and light absorption. As a consequence, it implies the necessity of using buffer layers, in such applications such as frequency conversion in periodically poled systems [149], in which otherwise it is unnecessary. Moreover, the poling fields are limited due to the microcircuits connected with the point effect. This leads also to unwanted and prohibitory increase of the optical propagation losses.

In all techniques the chromophore orientation is frozen by cooling the poled film to the room temperature under the applied external field (Fig. 24) or by thermal or photocrosslinking during the poling procedure. In the case of thermal crosslinking it is important to tightly control the poling temperature and to in-

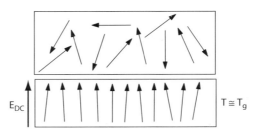

Fig. 24. Static field poling mechanism of dipolar molecules. Before applying electric field the dipole moments are randomly distributed. At higher temperatures the dipole moments of NLO chromophores are mobile and orient in the direction of the applied external field. The orientation is frozen by cooling to low temperature and/or photo- or thermal crosslinking

crease it simultaneously with the increasing crosslinking rate, as the glass transition temperature is also increasing. Shining with UV light during poling usually leads to an unwanted and uncontrollable increase of the poling current; therefore, it is recommended to make poling alternatively with photocrosslinking. Similarly, as in the case of thermal cross-linking, the glass transition temperature also increases. Therefore, it is also important to increase the poling temperature during the photocrosslinking process.

4.1.1
Establishment of the Axial Order

As already mentioned, the poling field orients the molecule dipole moment in its direction, which (at least in situations depicted in Figs. 21 and 22) is usually perpendicular to the thin film surface. As the poled chromophores are strongly anisotropic, such a change of the orientation of dipole moments may be well monitored by the observation of the thin film optical absorption spectrum [150, 151]. Figure 25 shows the temporal variation of the optical absorption spectrum of a side-chain PMMA-DR 1 polymer. Two observations may be made from the temporal behavior of the linear absorption spectrum as shown in Fig. 26:

(i) The optical density of poled film decreases during the poling process and increases in the following days after stopping it. The first effect is due to the already-

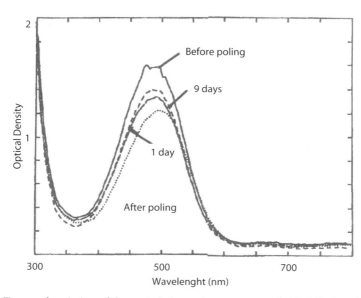

Fig. 25. Temporal variation of the optical absorption spectrum of a PMMA doped with DR 1 chromophore [151]

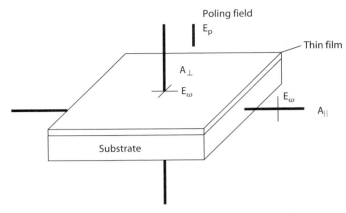

Fig. 26. Experimental geometry for measurement of the variation of linear absorption spectrum due to poling [151]

mentioned change of the orientation of the dipolar transition moments, while the second is due to the relaxation of the induced order (return of chromophores to the initial orientation).

The decrease of the optical absorption spectrum is due to the incident light polarization in the film plane, whereas molecules orient, under poling, in the direction perpendicular to the thin film surface. From this variation one can determine the order parameter $<P_2>$ using the following equation:

$$\varphi = \frac{A_{||} - A_\perp}{A_{||} + 2A_\perp} = 1 - \frac{A_\perp}{A_0} \tag{25}$$

where $A_{||}$ and $A_\perp$ ($A_{||} + 2A_\perp = 3A_0$) are absorbances (optical densities) measured with the incident light polarization parallel and perpendicular to the poling field direction, respectively (Fig. 26). Typically, the order parameters $<P_2>$ take the values between 0.1 and 0.3 for isotropic polymers, 0.5–0.6 for nematic liquid crystals, and 0.9–1 for smectic liquid crystals, respectively. It is worth noting that the order parameter $<P_2>$ determined in this way does not discriminate between polar and axial order. The only way to get such a discrimination is by using NLO techniques which are sensitive to the polar order.

(ii) The wavelength corresponding to the maximum absorption spectrum shifts towards larger values. This is due to the Stark effect as poled polymers behave like ferroelectrics with a large (a few MV cm^{-1}) poling fields. Shifts towards either higher or lower wavelengths are possible. (iii) Indeed the Stark shift Δ depends on the sign of the difference $\Delta\mu = \mu_{11} - \mu_{00}$ between molecule dipole moments in excited (1) and in fundamental (0) states

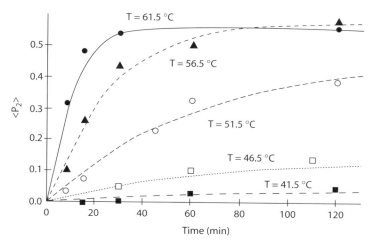

Fig. 27. Temporal variation of the order parameter $<P_2>$ of a polyacrylate PA3 ($T_g = 51.5\,°C$, $T_{NI} = 90\,°C$) at different poling temperatures [169]

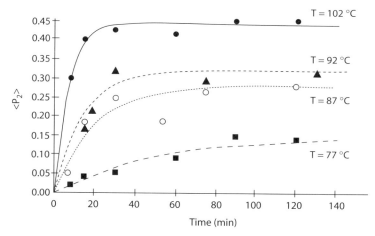

Fig. 28. Temporal variation of the order parameter $<P_2>$ of a polymethacrylate PMA3 ($T_g = 97\,°C$) at different poling temperatures [169]

$$\Delta \propto \frac{\Delta \mu E_p}{\eta c} \qquad (26)$$

where c is the light speed and $\eta = h/2\pi$, where h is Planck's constant.

The build up of the order parameter with poling time t may be well described by a monoexponential function [152, 153]:

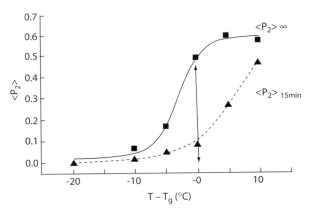

Fig. 29. Temperature variation of the saturation value of the order parameter $<P_2>_\infty$ for a polyacrylate PA3 ($T_g = 51.5$ °C, $T_{NI} = 90$ °C). The same after 15 min of poling ($<P_2>_{15\,min}$) is also shown [152]

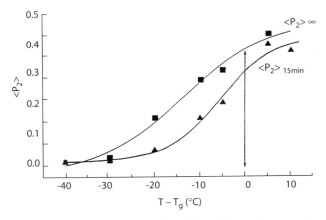

Fig. 30. Temperature variation of the saturation value of the order parameter $<P_2>_\infty$ for a PMA3 ($T_g = 97$ °C). The same after 15 min of poling ($<P_2>_{15\,min}$) is also shown [152]

$$\langle P_2 \rangle = \langle P_2 \rangle_\infty \left[1 - e^{-\frac{t}{\tau}} \right] \tag{27}$$

where τ is the time constant and $<P_2>_\infty$ is the plateau (saturation) value, both depending on the poling temperature (Figs. 27 and 28). Figures 27 and 28 show the poling kinetics for a side-chain liquid crystalline polymer (polyacrylate, PA3) and for an isotropic side-chain polymer [poly(methyl methacrylate), PMA3], both functionalized with a cyanobiphenyl chromophore. It is observed that one gets higher order parameters in the case of side-chain liquid crystalline polymers. Also the poling kinetics are quite different in these two cases. In the case of the liquid crystalline polymer the plateau value of the order parameter $<P_2>_\infty$ undergoes a percolation type behavior when approaching the glass transition temperature (Fig. 29), while in the second case it shows an almost monotonic variation (Fig. 30). For comparison, the temperature variation of the order parameter after 15 min of poling ($<P_2>_{15\,min}$) is also shown in both cases.

4.1.2
Build-Up of the Polar Order

As already mentioned, the only techniques sensitive to the polar order are even order nonlinear optical techniques such as the already-described second harmonic generation and linear electro-optic effect (cf. Chapter 2). The first technique offers a high sensitivity to the fast electronic contributions to $\chi^{(2)}$ susceptibility and is widely used. As already mentioned, it also gives the opportunity to study the kinetics of the poling by in situ measurements [152].

The build-up of the polar order, imaged by the growth of the SHG intensity, can be described by a triexponential function (Fig. 31):

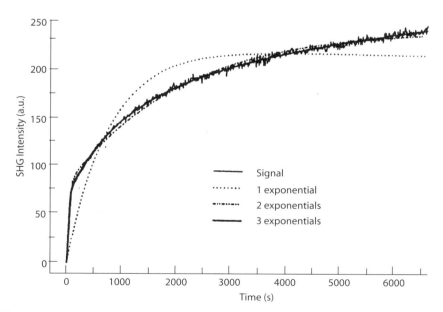

Fig. 31. Temporal variation of the SHG intensity during in situ poling process. *Points* show experimental data whereas *dashed, broken,* and *solid lines* show least square fit with a mono -, two-, and three-exponential equations, respectively [152]

$$\frac{1}{l}\left[\frac{I^{p}_{2\omega}}{I^{r}_{2\omega}}\right]^{1/2} = P_1\left(1-e^{-t/\tau_1}\right)+P_2\left(1-e^{-t/\tau_2}\right)+P_3\left(1-e^{-t/\tau_3}\right) \tag{28}$$

where l is the thin film thickness, τ_{1-3} are time constants of different orientation processes, and P_{1-3} are rates of different mechanisms taking part in the molecule orientation process, respectively.

The sum of the poling saturation limit values for different processes ($P = P_1 + P_2 + P_3$) represents the overall poling efficiency under the experimental conditions: temperature, geometry, poling field, atmosphere.

The superscripts p and r in Eq. (28) refer to polymer thin film and reference harmonic intensities. The reference, with simultaneous measurements of SHG intensity from a poled, stabilized thin film, was used to correct for eventual laser intensity fluctuations. The fits of the measured temporal growth of SHG intensity with mono-, bi-, and triexponential functions are shown in Fig. 31. The best agreement is obtained when using a triexponential function (Eq. (28)). The time constants τ_1, τ_2, and τ_3 depend on the polymer, temperature, and the chromophore itself.

The different time constants describing the growth of the polar order during poling may be attributed to different rotational mechanisms linked to the chromophore and the polymer (Fig. 32), as was discussed by Noël and Kajzar [154]. The initial rapid growth of SHG signal is characterized by a time constant τ_1 of a few

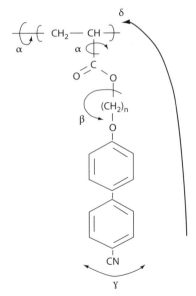

Fig. 32. Different modes intervening in orientation and relaxation processes [169]

seconds. This corresponds to the short period of time required to allow the surface charge density to build up in the corona poling set-up. During that time the system can be considered as an isotropic medium. The molecular energy depends only on the dipole orientation with respect to the directing electric field. Accordingly, this rapid step-growth of the SHG signal can be attributed to the fast reorientation response of the uncorrelated polar side-chain chromophores which are free to rotate. This process is mainly governed by local non-cooperative motions of the side-groups (sub-T_g β-relaxation modes, Fig. 32). Its contribution to the SHG signal depends on three factors:

(i)   size of the local free volume around each chromophore
(ii)   degree of decoupling between the chromophore and the polymer backbone
(iii) poling temperature, determining the thermal fluctuation energy.

The subsequent growth of the SHG signal (described by τ_2) is due to an improved alignment of chromophores under the quickly established, constant electric field and to the development of nematic order in the case of liquid crystals. The energy of the system depends not only on the orientation of molecule dipoles relative to the poling direction, but also on the tendency towards mutual axial alignment experienced by the active chromophore groups. For LC polymers, in which the mesogenic groups are attached to the polymer backbone via short spacers only, a partial decoupling of the main- and side-chain motions is obtained. Hence, the side-chain order influences the polymer chain contour. This results in the distortion of the polymer backbone from its normal random-coil conformation. The orienting electric field, while leading to an alignment of the active mesogenic groups, also causes the backbone units to change their conformation locally to adapt to the anisotropic orientation distribution of the poled chromophores. Therefore, the polarization growth process with the time constant τ_2 can be associated with the reorientation motions of both the polymer backbone chain segments (α-relaxation mode, Fig. 32) and the side-chains. Its maximum contribution depends on the polymer segmental mobility and the degree of decoupling between the polymer segments and the chromophores.

The last SHG signal growth process, described by time constant τ_3, not observed with amorphous isotropic polymers, may be associated with the ultraslow reorientation motions of the chromophores about the small axis. These motions give rise to the so-called δ-relaxation mode (Fig. 32), controlled by free volume effects. However, this process is highly cooperative and requires more free volume than that characteristic of the main chain segmental motions. For PMA3, the thermal stability range of the mesophase developed under the poling field is rather small and this slow step-growth is absent from the SHG signal as soon as the poling temperature is a few degrees above T_g. For PA3, at a critical poling temperature ($T_g + 25\ °C$), the poling mechanism 2 and 3 tends to merge. This effect could result from the coupling between the different relaxation modes.

4.2
Photoassisted Poling

Dumont and coworkers [136, 155, 156] have observed that shining doped (or functionalized) polymer thin films with noncentrosymmetric dipolar chromophores, induces a significant increase of electro-optic coefficient in the chromophore absorption band, corresponding to a better, polar orientation of chromophores. The measurements have been done by using the attenuated total reflection technique, and the optical field polarization was perpendicular to the applied low-frequency external electric field to the thin film (Fig. 33). A better stability of induced orientation was observed in the case of functionalized polymers than in guest-host system, as is usually the case with the static field poled polymers. The chromophores orient with dipolar moments perpendicular to the optical field (and parallel to the applied static (or low frequency) field. As will be discussed later, the chromophore orientation undergoes a *trans-cis* isomerization process (Fig. 34).

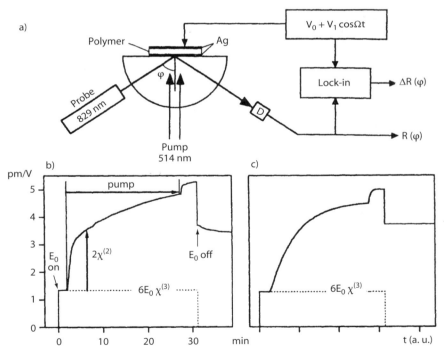

Fig. 33a–c. Attenuated total reflection set-up (**a**) and temporal growth of the electro-optic coefficient as function of laser illumination at 514.5 nm: (**b**) experiment, (**c**) theory (courtesy of M. Dumont)

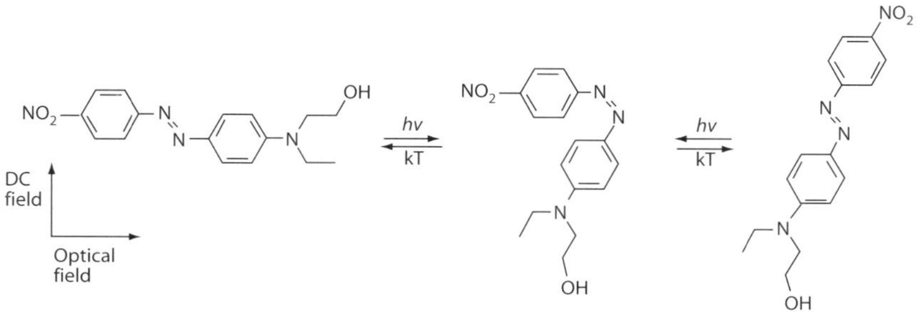

Fig. 34. Light-induced reorientation of the DR 1 molecule in the presence of the static field and photoisomerization process. From almost parallel orientation to the exciting optical field the molecule dipole moments reorient to an almost perpendicular direction

Figure 33 shows the experimental set-up used by Dumont and coworkers together with the observed and calculated dependence of the electro-optic coefficient on the external light illumination (green argon laser line). The studied, poled film is placed between two electrodes – one of them a silver electrode deposited on the side of a prism. The condition to excite surface polaritons in thin film depends on its refractive index and thickness as well as on the refractive indices of electrode material and prism. By varying the incidence angle it is possible to satisfy this condition, which is observed as a dip in the intensity of reflected beam. By varying the refractive index of the thin film with the applied voltage, one changes the resonance conditions (or in other words the coupling angle). The technique is very sensitive to thin film thickness and refractive index variation. As the variation of refractive index depends on the thin film electro-optic coefficient the technique serves to determine it with a high precision. The variation of refractive indices is measured with a lock-in amplifier. By shining the studied film with a He-Ne laser emitting at 632.8 nm or with an argon laser at 514.5 nm parallel to the poling field (optical electric field perpendicular to it) Dumont et al. observed a noticeable increase of the measured electro-optic coefficient (Fig. 33). The effect was significantly larger in the case of grafted polymer than with a guest-host system. Switching off the light source leads to a decrease of electro-optic coefficient, which is connected with the decrease of the polar order. The relaxation of polar order is faster in guest-host systems than in side-chain polymers, as it is usually observed.

A possible explanation of this effect is shown schematically in Fig. 34. The active molecule, which is the DR 1 chromophore, undergoes the *trans-cis* isomerization under the light illumination. The double N = N bond is flexible in the excited state and through rotation or translation, the molecule may change configuration from *trans* to *cis* form. This process is very fast in liquids [157,158] and is significantly slower in the solid state [159]; it has been estimated to take about 150 ns for a grafted polymer. The inverse transformation (although sometimes the process is

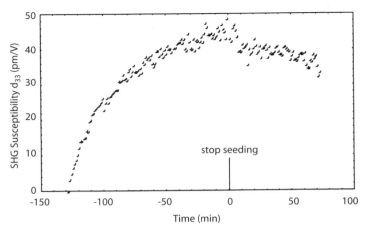

Fig. 35. Experimental growth and decay of the second-order NLO susceptibility

irreversible) from *cis* to *trans* form may go only through nonradiative channels and is very slow (of the order of a few seconds in solids). Molecules can return to the previous configuration, but will again be excited by incoming light. Thus, the only stable position will be obtained if molecule orients with the dipole moment perpendicular to the incident light polarization, thus parallel to the applied electric field. As a consequence, the light-induced molecular re-orientation will lead to an increase of the electro-optic coefficient as was observed experimentally by Dumont and coworkers [136, 155, 156, 160]. By using a simple rate equation for the *trans-cis* isomerization process Dumont [160] described the observed temporal behavior of the electro-optic coefficient during the attenuated total reflection measurements (solid lines in Fig. 35).

4.3
All Optical Poling

Charra et al. [137] and Kajzar et al. [159] observed formation of a polar orientation of chromophores in nondegenerate (pump beams at 1064 nm, probe beam at 532 nm) phase conjugate experiments performed on a copolymer of PMMA with DR 1 chromophore. The observed signal at 532 nm rose slowly with time, up to a saturation value. A spontaneous SHG was observed after switching off the probe beam, with a fast relaxation component at the beginning [159]. The maximum second-order nonlinear optical susceptibility value obtained in these experiments was 3 pm V^{-1} [137]. This experiment showed that by using purely optical fields one could obtain a polar orientation of chromophores in a functionalized or a doped polymer film.

A significantly larger $\chi^{(2)}$ value was obtained in seeding geometry by using two collinear picosecond beams at 1064 nm and 532 nm. The chromophore orientation was done by using a pulsed picosecond Nd:YAG laser delivering both fundamental (1064 nm) and harmonic (532 nm) wavelengths at 10 Hz repetition rate. Energies were 500 µJ and 2.5 µJ, respectively at 1064 nm and 532 nm, and the beam diameter was 2 mm at the sample location. With the same polymer as that used in the four-wave mixing geometry, the values of d_{pp} susceptibility close to that obtained in corona poling were measured [161, 162]. An example of the poling process for a PMMA-DR 1 of 65/35 mol % copolymer is shown in Fig. 35. The growth of the polar order is much slower than in the case of static field poling, where it takes place during about three minutes at temperatures close to the polymer glass transition temperature. Here, the poling is done at room temperature. After stopping the seeding procedure, a relaxation of the induced order is observed, as is also the case with static field poling. This seeding geometry had been used for optical poling of glass fibers, where an efficient SHG was also obtained after the seeding procedure [163, 164].

The mechanism responsible for the creation of $\chi^{(2)}$ grating in glass fibers was proposed originally by Baranova and Zeldovich [165] in terms of a polychromatic interference of input fields at ω and 2ω frequencies, leading to a non-zero temporal average poling field:

$$\left\langle \left[E_{2\omega} \cos 2\omega t \left(E\omega \cos(\omega t + \phi) \right)^2 + c.c \right] \right\rangle_t = E_{2\omega} E_\omega^2 \cos(\Delta\phi) \qquad (29)$$

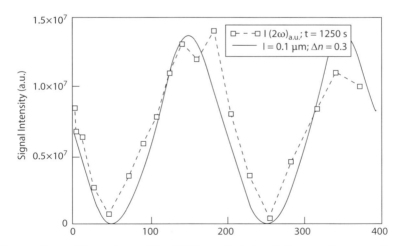

Fig. 36. Experimental dependence of the SHG intensity induced after 20 min preparation time, on the relative phase $\Delta\Phi$ of the ω and 2ω beams. *Solid line* corresponds to a theoretical dependence with $\Delta n = n(2\omega) - n(\omega) = 0.3$. Sample was 0.1 µm thick with 0.3-optical density at 532 nm [161]

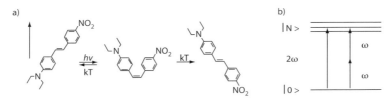

Fig. 37a,b. Schematic view of the orientation mechanism in all optical poling with azo dye molecules (**a**) and electron transition diagram between fundamental and excited singlet states (**b**). In the case of azo dyes, the *trans-cis* isomerization is achieved through both one- and two-photon excitations [144]

where $\Delta\phi$ is the relative phase difference between E_ω and $E_{2\omega}$ fields. Such a dependence was indeed verified, as shown in Fig. 36. However the microscopic mechanism of all optical poling in polymers is different from that in glass fibers, in which the color centers and defects are at the origin of the created polarization. Similarly as in photoassisted poling and in the case of azo dyes, the polar orientation of dipole moments can be explained by the *trans-cis* isomerization. Contrariwise to the photoassisted poling, in which this isomerisation is induced by a one-photon transitions (Fig. 34), in all optical poling this is excited simultaneously by two photons with frequency ω and one photon with frequency 2ω (Fig. 37).

As already mentioned, the azo dyes used strongly absorb light if the exciting optical field is parallel to the dipole transition moment. The double N = N bond in the excited state is mobile and the molecule changes conformation to the *cis* form, with a smaller volume and with increased mobility. Subsequently, the molecule relaxes slowly to the *trans* form through the non-radiative channels. In photoassisted poling a stable orientation of chromophores is achieved when the dipole transition moment is perpendicular to the optical, exciting field. A given polar orientation is imposed by the applied DC field. In all optical poling, a stable chromophore orientation is obtained when the dipole moment is directed oppositely to the exciting field due to the existence of a non-zero temporal average "poling" field (Eq. (29)). There is a fundamental difference between photoassisted poling and all optical poling in the chromophore orientation obtained. In the first case, chromophores orient perpendicularly to the optical field, whereas in the second they are parallel (to the resulting $<E^3>$poling field).

5
Statistical Orientation Models

For noninteracting dipole moments and for a single crystal there exists a straight relationship between the microscopic (β_{ijk}) and macroscopic $\chi^{(2)}_{IJK}$ hyperpolarizabilities. This is obtained by the transformation of the corresponding tensor components from the molecular to the laboratory reference frame:

$$\chi_{IJK}^{(2)}\left(-\omega_3;\omega_1,\omega_2\right)=$$

$$\sum_n N^{(n)} \sum_{ijk} f_i^{(n)\omega_1} f_j^{(n)\omega_2} f_k^{(n)\omega_3} a_{iI}^{(n)} a_{jJ}^{(n)} a_{kK}^{(n)} \beta_{ijk}^{(n)}\left(-\omega_3;\omega_1,\omega_2\right) \tag{30}$$

where $N^{(n)}$ is the number density of molecular species (n), a_{iI} are Wigner's rotation matrices, ω_{1-3} are frequencies of interacting photons (a static field is considered as a photon with frequency 0) and $f^{(n)\omega}$ is the local field factor at frequency ω, as given by Eq. (7). This relationship is not more valid for partially ordered systems only, like poled polymers for which the microscopic second-order NLO susceptibility is given by:

$$\chi_{IJK}^{(2)}\left(-\omega_3;\omega_1,\omega_2\right)= NF\left\langle\beta_{ijk}\left(-\omega_3;\omega_1,\omega_2\right)\right\rangle_{IJK} \tag{31}$$

where for the sake of simplicity we assumed only one active species with density N, $<>$ in Eq. (31) is the orientation average of β hyperpolarizability, and F is the global field factor (Eq. (6) with corresponding frequencies).

As already mentioned, for poled polymers with point symmetry ∞ mm and taking into account the Kleinman's conditions there are two non-zero $\chi^{(2)}$ tensor components: diagonal $\chi_{zzz}^{(2)}$ and off diagonal $\chi_{xxz}^{(2)}$, where the z-axis is parallel to the poling direction. Further simplification comes by considering only the one-dimensional charge-transfer molecules, as is the case here, with enhanced β_{zzz} component in the charge-transfer direction z (in the molecular reference frame). In that case we have the following relationships between the macroscopic and the microscopic quantities (for the sake of simplicity we assume the azimuthal symmetry as it is the case with poled polymers):

$$\chi_{ZZZ}^{(2)}\left(-\omega_3;\omega_1,\omega_2\right)= NF\beta_{zzz}\left(-\omega_3;\omega_1,\omega_2\right)\left\langle\cos^3\Theta\right\rangle \tag{32}$$

and

$$\chi_{XXZ}^{(2)}\left(-\omega_3;\omega_1,\omega_2\right)=\frac{1}{2}NF\beta_{zzz}\left(-\omega_3;\omega_1,\omega_2\right)\left\langle\sin^2\Theta\cos\Theta\right\rangle \tag{33}$$

where Θ is the angle between the poling field and the molecular (charge transfer) axis (Fig. 38).

The orientational averages appearing in Eqs. (32) and (33) can be obtained by introducing the orientation distribution function $G(\Theta)$, giving the probability of finding the molecular axis in the Θ direction (Fig. 38). In that case the corresponding orientation averages are given by:

$$\left\langle\cos^3\Theta\right\rangle=\frac{2\pi}{N_A}\int_0^\pi G(\Theta)\cos^3\Theta\sin\Theta d\Theta \tag{34}$$

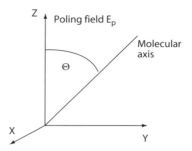

Fig. 38. Molecular axis orientation with respect to the poling field E_p

and

$$\left\langle \cos\Theta \sin^2\Theta \right\rangle = \frac{2\pi}{N_A} \int\limits_0^\pi G(\Theta)\cos\Theta \sin^3\Theta d\Theta \tag{35}$$

where the normalization factor is given by:

$$N_A = 2\pi \int\limits_0^\pi G(\Theta)\sin\Theta d\Theta \tag{36}$$

At higher temperatures (above or slightly below the room temperature), when molecules may rotate freely (free gas model) one uses the Gibbs-Boltzmann distribution function for $G(\theta)$ given by:

$$G(\theta, E_p) = e^{\frac{U(E_p)}{kT}} \tag{37}$$

where E_p is the poling field. For dipolar molecules, the ordering energy U is given by:

$$U(E_p) = \mu^* E_p \cos\theta + \frac{1}{2}\left(\alpha_\| - \alpha_\perp\right)E_p^2 \cos^2\theta = U_1(E_p) + U_2(E_p) \tag{38}$$

where the first term on the right-hand side is the electric field-dipole moment interaction energy and the second term describes interaction energy of the induced dipole moment with electric field. $\alpha_\|$ and $\alpha_\perp$ are molecular hyperpolarizabilities parallel and perpendicular to the charge transfer direction, respectively, and μ^* is the ground-state dipole moment of the molecule, corrected for the local field. Neglecting the induced dipole moment interaction energy with the poling field (isotropic model), for the second-order NLO susceptibility components, one gets the following expressions [166]:

$$\chi_{ZZZ}^{(2)}\left(-\omega_3;\omega_1,\omega_2\right) = NF\beta_{zzz}\left(-\omega_3;\omega_1,\omega_2\right)L_3(x) \tag{39}$$

and

$$\chi_{XXZ}^{(2)}\left(-\omega_3;\omega_1,\omega_2\right) = NF\beta_{zzz}\left(-\omega_3;\omega_1,\omega_2\right)\left(L_1(x)-L_3(x)\right) \tag{40}$$

where

$$x = \frac{\overset{*}{\mu}E_p}{kT} \tag{41}$$

is the ratio of ordering energy to the thermal randomization energy. $L_n(x)$ in Eqs. (39) and (40) is the Langevin function given by:

$$L_n(x) = \frac{\int \cos^n \Theta e^{x\cos\Theta}\sin\Theta d\Theta}{\int e^{x\cos\Theta}\sin\Theta d\Theta} \tag{42}$$

with

$$L_1(x) = \coth x - \frac{1}{x} \tag{43}$$

$$L_2(x) = 1 + \frac{2}{x^2} - \frac{2}{x}\coth x \tag{44}$$

$$L_3(x) = \left(1 + \frac{6}{x^2}\right)\coth x - \frac{3}{x}\left(1 + \frac{2}{x^2}\right) \tag{45}$$

It is straightforward to express the order parameter $<P_2>$ in terms of Langevin function $L_2(x)$ [167]:

$$\langle P_2 \rangle = \langle P_2(\cos\Theta) \rangle = \frac{1}{2}\langle 3\cos^2\Theta - 1\rangle = \frac{3}{2}(L_2(x)-1) \cong \frac{x^2}{15} \text{ for } x \leq 1 \tag{46}$$

From the knowledge of the order parameter from the measurement of the variation of the optical absorption spectrum due to poling, one can estimate the x parameter (Eq. (43)) intervening in the above developments. For poled polymers one can use also another alternative description of linear and nonlinear optical susceptibilities by expanding the orientation distribution function in the series of Legendre polynomials, where the expansion coefficients are order parameters:

$$G(\theta) = \sum_{n=0}^{\infty} \frac{2n+1}{2}\langle P_n(\cos\theta)\rangle P_n(\cos\theta) \tag{47}$$

which can be expressed in terms of the modified spherical Bessel functions [168]

$$\left\langle P_n(\cos\theta)\right\rangle = \frac{i_n(x)}{i_0(x)} \tag{48}$$

The following recurrent relations obey the modified spherical Bessel functions:

$$i_{n-1}(x) - i_{n+1}(x) = \frac{2n+1}{x} i_n(x) \tag{49}$$

where

$$i_0(x) = \frac{\sinh x}{x} \tag{50}$$

Similarly, for Legendre polynomials we have

$$\left\langle P_{n-1}(\cos\Theta)\right\rangle - \left\langle P_{n+1}(\cos\Theta)\right\rangle = \frac{2n+1}{x}\left\langle P_n(\cos\Theta)\right\rangle \tag{51}$$

with

$$\left\langle P_0(\cos\Theta)\right\rangle = 1 \tag{52}$$

$$\left\langle P_1(\cos\Theta)\right\rangle = \left\langle\cos\Theta\right\rangle \tag{53}$$

$$\left\langle P_2(\cos\Theta)\right\rangle = \frac{1}{2}\left\langle 3\cos^2\Theta - 1\right\rangle \tag{54}$$

$$\left\langle P_3(\cos\Theta)\right\rangle = \frac{1}{2}\left\langle\left(5\cos^3\Theta - 3\cos\Theta\right)\right\rangle \tag{55}$$

and

$$\left\langle P_4\right\rangle = \frac{1}{8}\left\langle 35\cos^4\Theta - 30\cos^2\Theta + 3\right\rangle \tag{56}$$

The odd parameters $<P_{2k+1}>$ characterize the polar order, while the even order parameters characterize the axial order.

Introducing the developments (Eqs. (43)–(45)) into Eqs. (39) and (40)) one can calculate the effect induced by the poling field via the second-order NLO susceptibility tensor components $\chi^{(2)}_{zzz}$ and $\chi^{(2)}_{xxz}$. Both are depicted in Fig. 39 as a function of the x parameter (Eq. (41)). It is clearly seen that the poling process, with poling field directed perpendicular to the thin film surface, creates noncentrosymmetry in thin film with both components increasing for small x values (Eq. (41)). The off diagonal $\chi^{(2)}_{xxz}$ component reaches a maximum for $x \approx 3$ and thereafter decreases, while the diagonal component $\chi^{(2)}_{zzz}$ component increases monotonically

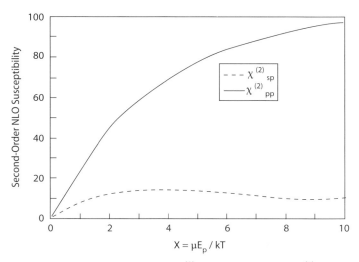

Fig. 39. Calculated variation of the diagonal ($\chi^{(2)}_{ZZZ}$) and off diagonal ($\chi^{(2)}_{XXZ}$) tensor components with $x = \mu E_p / kT$

to a saturation value corresponding to a perfect alignment at infinite electric field (all dipole moments parallel). We note here that at the same time the poling process creates optical birefringence in the thin film with an increase of extraordinary and a decrease of ordinary index of refraction, as can also be seen from the optical absorption variation through Kramers-Kronog relations. Both $\chi^{(2)}$ tensor components (Eqs. (32) and (33)) can be expressed in terms of the polar order parameters $<P_1>$ and $<P_3>$ through Eqs. (53) and (55) yielding:

$$\chi^{(2)}_{ZZZ}\left(-\omega_3;\omega_1,\omega_2\right) = NF\beta_{zzz}\left(-\omega_3;\omega_1,\omega_2\right)\left[\frac{3}{5}\left\langle P_1\left(\cos\theta\right)\right\rangle + \frac{2}{5}\left\langle P_3\left(\cos\theta\right)\right\rangle\right] \quad (57)$$

and

$$\chi^{(2)}_{ZZZ}\left(-\omega_3;\omega_1,\omega_2\right) = NF\beta_{zzz}\left(-\omega_3;\omega_1,\omega_2\right)\left[\frac{1}{5}\left\langle P_1\left(\cos\theta\right)\right\rangle - \frac{1}{5}\left\langle P_3\left(\cos\theta\right)\right\rangle\right] \quad (58)$$

The efficiency of the polar ordering is given by the ratio

$$a = \frac{\chi^{(2)}_{ZZZ}}{\chi^{(2)}_{XXZ}} \quad (59)$$

which varies between 1 and ∞. The last value is reached for perfectly ordered structures (all dipole moments pointing in the same direction). The parameter $a = 3$ for a free electron gas (isotropic model). For side-chain liquid crystalline polymers a values as high as 18 have been obtained [169].

6
Relaxation Processes

One of the important parameters determining practical applicability of poled polymers is the stability of the induced polar orientation. This is studied usually through the temporal behavior of the $\chi^{(2)}$ susceptibility or of the electro-optic coefficient r at elevated temperatures. The studies of the relaxation of the induced polar order were almost done for the static field poled polymers. The temporal decay of the $\chi^{(2)}$ susceptibility with t is usually described by the Kohlrausch [170]-Williams-Watt [171] (KWW) stretched exponential function:

$$\chi^{(2)}(t) = \chi^{(2)}(t = 0)e^{-(t/1)^{\beta}} \tag{60}$$

where τ is the temperature-dependent relaxation time constant and β ($0 < \beta < 1$) describes the width of relaxation (departure from a monoexponential behavior).

The relaxation of poled polymers (isotropic or liquid crystalline) may be also described by a biexponential function:

$$\frac{1}{l}\left[\frac{I_{2\omega}^{p}(t)}{I_{2\omega}^{r}(t)}\right]^{\frac{1}{2}} = R_1 e^{-t/\tau_1} + R_2 e^{-t/\tau_2} + C \tag{61}$$

where l is the thin film thickness, $I_{2\omega}^{p(r)}$ is the second harmonic intensity of the studied thin film (p) and of reference (r), respectively. The constant C in Eq. (61) characterizes the residual orientation at the experimental time scale and is an important parameter for practical applications, determining the long-term stability of

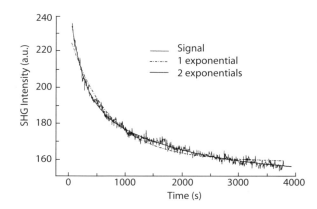

Fig. 40. Temporal decay of SHG intensity at elevated temperature for an oriented polyacrylate functionalized with cyanobiphenyl chromophore. *Dash-dotted* and *solid lines* have been computed using one and biexponential functions, respectively [154]

induced order. R_1 and R_2 in Eq. (61) are relaxation rates and τ_1 and τ_2 are relaxation times of different processes contributing to the molecular disorientation, respectively. All these parameters depend on the measurement temperature. The closer the glass transition temperature is to the measurement temperature, the smaller the time constants τ and the larger the relaxation rates. This behavior is true for both isotropic and liquid crystalline polymers, as was observed by Dantas de Morais et al. [152]. Figure 40 shows an example of such a biexponential fit of $\chi^{(2)}$ decay curve.

7
Light-Induced Depoling

Large et al. [172] and Combellas et al. [173, 174] observed that shining a light on a poled polymer film within its absorption band leads to a rapid destruction of the polar order. This process is perfectly reversible if done when the film is subjected to a static external field (corona or electrode). The experiments were performed by an in situ SHG technique. Figures 41 and 42 show the temporal behavior of SHG intensity in the case of a polymer film functionalized with PMMA photoisomerizable chromophore (DR 1) and a zwitterionic molecule dissolved in PMMA matrix, respectively. In both cases one observes a disappearance of SHG signal when shining in the chromophore absorption band. The SHG signal reappears when the disorienting laser beam is switched off. The amount of disoriented chromophores depends on the number of absorbed photons. The disorientation is complete when the number of absorbed photons exceeds the number of chromophores. While in the case of the DR 1 chromophore the SHG intensity is almost constant after each switch-off of the disorienting laser beam, it increases gradually in the case of the

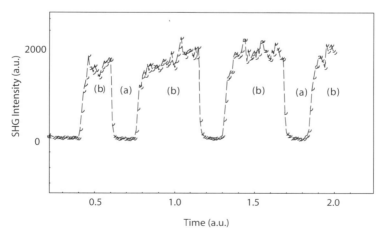

Fig. 41a,b. Temporal variation of the measured in situ SHG intensity under (**a**) and without (**b**) thin film illumination in its absorption band with a He-Ne laser [172]

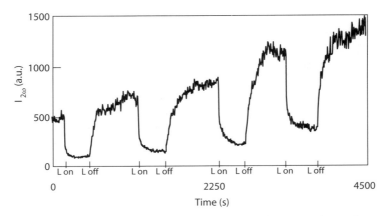

Fig. 42. Temporal variation of the second harmonic intensity from guest-host polymer doped with zwitterionic chromophore film and submitted to the illumination periods in the chromophore absorption band (541 nm) between t = L on and t = L off and during the corona poling [173, 174]

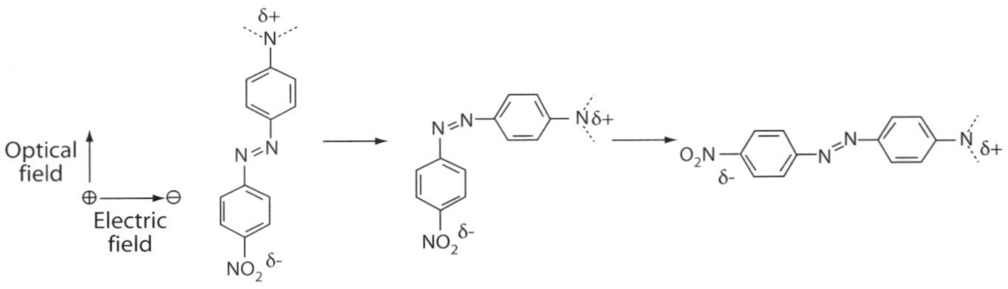

Fig. 43. Proposed mechanism of light-induced depoling process with photoisomerizable molecules

zwitterionic molecule. It shows that the depoling process is different in these two cases. In the former case it can be explained by the *cis-trans* isomerization process (Fig. 43).

In the optical depoling process the exciting optical field is perpendicular to the poling static field. As already mentioned, the quasi-one-dimensional azo dyes are strongly anisotropic and absorb light when the optical field points in the transition dipole moment direction. Excited molecules change conformation from *trans* to *cis* form, with smaller volume and higher mobility. The return to the equilibrium "*trans*" form is accompanied by a rotation of the electron acceptor group, resulting in an orientation anti-parallel to the electric field (perpendicular to the optical field). There is therefore a light-induced reduction in the net polar order.

The modulation of the $\chi^{(2)}$ susceptibility, which is monitored by the variation of SHG intensity, is associated with the change of the medium refractive index Δn given by:

$$\Delta n = -\frac{n^3 \Delta r E}{2} \tag{62}$$

where $\Delta r \propto \Delta \chi^{(2)} \propto \Delta I_{2\omega}$, E is the poling electric field, and n is the refractive at the operation wavelength. The change in the refractive index can be calculated by using typical values for the above quantities. By assuming easily attainable $\Delta r = 10$ pm V^{-1}, $E_p = 2$ MV cm^{-1}, and $n = 1.6$, one gets a value for Δn of 0.01, which is sufficient to modify the waveguiding conditions and may find numerous applications.

What is also very interesting from the practical point of view is that such an alternate illumination of the zwitterionic chromophore during the corona poling leads to a significant increase of $\chi^{(2)}$ susceptibility. The observed increase of $I_{2\omega}$ intensity by a factor of three (Fig. 42) corresponds to an improvement of $\chi^{(2)}$ susceptibility (and correspondingly of the E-O coefficient r) by 70%. Also, the temporal stability of the induced noncentrosymmetric order in alternately illuminated thin films is significantly better than in untreated films (an increase of the decay time constant by a factor of 5 [174]).

8
Applications of Second-Order NLO Polymers

The noncentrosymmetric materials described in this paper target a large class of practical applications such as frequency doubling (blue conversion for data storage, microlithography, medicine, biology), tunable light sources (optical parametric oscillators), electro-optical modulation for high-rate (tens to hundreds of GHz) signal transmission, terahertz electric pulse generation, not achievable by purely electric circuitry, etc. Some of these applications were already realized with poled polymers with different performances and will shortly be discussed and described in this chapter. For other realizations and more details, in particular concerning the electro-optic modulation, the reader is refered to the chapter by L. Dalton.

8.1
Frequency Doubling

The poled polymers were developed mainly for applications in waveguiding configuration which we will discuss shortly here. One of them is frequency doubling through the SHG process. Several techniques were proposed and developed for these purposes and polymeric waveguides. These are:

(i)   Cerenkov-type SHG generation [175–177]
(ii) Quasi-phase matching in periodically poled polymer films [178, 179, 149]
(iii) Counterpropagating beams SHG [180, 181]
(iv) Modal phase matching [182]

8.1.1
Cerenkov-Type SHG Generation

In the Cerenkov-type SHG the fundamental wave is guided in thin film, while the harmonic wave is outcoupled into the substrate (Fig. 44). It requires that the refractive indices of the substrate $n_s^{\omega(2\omega)}$ at ω and 2ω frequencies and the effective index of thin film n_{eff}^{ω} at ω frequency satisfy the following condition:

$$n_s^{2\omega} > n_{eff}^{\omega} > n_s^{\omega} \tag{63}$$

No phase matching is realized with this structure; thus, the conversion efficiency is limited. The main advantage is its simplicity and no severe requirements on the propagation properties at harmonic frequency (the film may be slightly absorbing); thus, higher $\chi^{(2)}$ values through resonance enhancement may be exploited (harmonic wavelength closer to the absorption band).

8.1.2
Quasi-Phase Matching in Periodically Poled Polymer Films

The structure of a periodically poled polymer is shown schematically in Fig. 45. By applying an alternatively directed poling DC field it is possible to create a structure

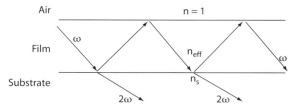

Fig. 44. Schematic representation of Cerenkov-type SHG in a polymeric waveguide

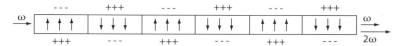

Fig. 45. Schematic representation of alternate quasi-phase-matched structure. *Arrows* shows orientation of dipolar moments (and consequently the orientation of the diagonal $\chi_{ZZZ}^{(2)}$ component)

composed from substructures with thickness corresponding to the coherence length in thin film. Due to the change of direction of $\chi^{(2)}_{ZZZ}$ component when passing from one substructure to the second one, there is no decrease in the intensity of SH wave, but free and bound waves still add constructively. Obviously such structure works at fixed wavelength only, although by rotating it slightly one can also convert wavelength close to the optimal conversion wavelength corresponding to the fundamental wave at normal incidence. Of course, due to the coupled intensity decrease with incidence angle, less conversion efficiency will be achieved.

8.1.3
Counter-Propagating Beams Second Harmonic Generation

Normandin et al. [181, 183–186] proposed a new geometry for SHG generation in poled polymer waveguides using counter-propagating beams at fundamental frequency, as shown schematically in Fig. 46. Two beams with ω frequency propagate in opposite direction, and in the area of their overlap 2ω beam is emitted perpendicular to the thin film surface with intensity [181]:

$$P_{2\omega} = A_{NL}P_+(0)P_-(L)\frac{L}{W} \tag{64}$$

where P_- and P_+ are input powers on both sides of the waveguide, respectively, W is the beam width, L is the propagation length, and

$$A_{NL} = \frac{1}{2}cn^{2\omega}\varepsilon_0|TS|^2 \tag{65}$$

where T is the effective transmission coefficient and S is the overlap integral

$$S = \frac{i\omega}{cn^{2\omega}}\int_{-\infty}^{0} d_{yyy}(-2\omega;\omega,\omega)E_y^2(x)e^{\frac{i2\omega n^{2\omega}}{c}x}dx \tag{66}$$

Although the conversion efficiency is not very high the technique can be used for beam demultiplexing in the two-wave mixing configuration. Due to the wave vector conservation requirement the direction of the sum frequency beam emission depends on the ratio of the incident beam frequencies.

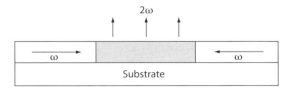

Fig. 46. Schematic representation of SHG with counter-propagating beams

8.1.4
Modal Phase Matching

In optical waveguides several modes may propagate, satisfying the eigen equation and whose number depends on the light polarization (TE or TM modes), the thin film thickness, its refractive index as well as on the substrate and superstrate (usually air) layer refractive indices. Different modes propagate with different velocities corresponding to different refractive indices. The modal phase matching can be realized if the effective refractive index of a fundamental (TE or TM) mode (f) is equal to that of a harmonic (TE or TM) mode (h)

$$n_{eff}^{f}(\omega) = n_{eff}^{h}(2\omega), f < h \tag{67}$$

The conversion efficiency is directly proportional to the overlap integral between fundamental and harmonic optical fields of given modes

$$S_{fh} = \frac{\left[\int\limits_{active\ layer} \left(E_f^{\omega}\right)^2 E_h^{2\omega} dz \right]^2}{\int\limits_{-\infty}^{\infty} \left(E_h^{2\omega}\right)^2 dz \left[\int\limits_{-\infty}^{\infty} \left(E_h^{2\omega}\right)^2 dz \right]^2} \tag{68}$$

Up to now, no significant conversion efficiencies $\eta = P_{2\omega}/P_{\omega}$, where P is power at given frequency, were reported. This is mainly due to the poor overlap integral, limited propagation length caused by the nonoptimized propagation properties of polymeric waveguides, and limited values of $\chi^{(2)}(-2\omega,\omega,\omega)$ susceptibility.

We note also that PM SHG was obtained with poled polymer thin films using a planar waveguide with corrugated grating [187].

8.2
Frequency Tuning

One of the important applications of second-order NLO materials is obtaining of tunable laser sources. Second harmonic generation or sum frequency generation systems lead to monochromatic sources. The optical parametric oscillators are based on the parametric generation of two waves with frequencies ω_s (signal) and ω_i (idler). In noncentrosymmetric materials an incident photon with frequency ω creates two photons satisfying the energy

$$\omega = \omega_s + \omega_i \tag{69}$$

and momentum conservation

$$k_\omega = k_{\omega_s} + k_{\omega_i} \tag{70}$$

Tunable OPOs were already demonstrated with different organic single crystals. Very recently Alshikh Khalil et al. [182] have shown the feasibility of an OPO in guided wave configuration with poled polymer film. The pump was at 532 nm and the signal was at 853 nm. Internal optical gain of 1 dB was obtained after propagating on 5 mm with the pump power of 1.5 kW.

8.3
Electro-Optic Modulation

Electro-optic modulation is another important field for applications of second-order NLO materials for signal transmission, signal processing, and optical interconnections. Poled polymers allow the making of electro-optic modulators in waveguiding configuration suitable for applications in integrated optics and for parallel treatment [188–191]. Moreover, these modulators can be integrated with the silicon technology, are easy and cheap to fabricate, and allow a transmission band over 400 GHz [192] with half-wave voltage less than 1 V [194–196]. Figure 47 shows an example of a polymeric Mach-Zehnder modulator for amplitude modulation. The light beam is split into two beams propagating in separate arms of the modulator. A multilayer structure (shown in insert of Fig. 47) is placed in one of the arms (although sometimes in both arms to get a better control of the relative phase of two beams) and depending on the applied electrode voltage, the phase of the propagating beam is varied. This is due to the variation of the refractive index of the active polymeric layer in which the beam propagates through the linear electro-optic effect (Eq. (62)), which induces a phase change of the propagating beam given by:

$$\Delta\varphi = \frac{\pi r n^3 L E}{\lambda} \tag{71}$$

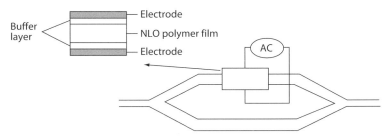

Fig. 47. Schematic representation of the structure of a Mach-Zehnder interferometer for amplitude modulation

where n is the polymer film refractive index, r is its electro-optic coefficient, L is the propagation length (waveguide length), λ is the operation wavelength, and E is the applied electric field. The resultant phase mismatch between two beams leads to the variation of the amplitude, and consequently of the intensity of two interfering beams at the output of the interferometer, which is controlled by the applied voltage to the electrodes. The minimum voltage necessary to create a phase mismatch between both beams equal to π is called the half-wave voltage V_π and is given by:

$$V_\pi = \frac{d\lambda}{rn^3 L\Gamma} \qquad (72)$$

where d is the waveguide thickness and Γ is a correction factor close to 1. The performances and in particular the bandwidth depends strongly on the design of electrodes. Actually, polymeric electro-optic modulators with a transmission band of 130 GHz have been reported [105, 106]. More details on polymeric Mach-Zehnder electro-optic modulators and their performances can be found in the chapter by Dalton [194].

Another type of intensity modulators was proposed and developed by Blau et al. [195]. It uses diffraction gratings (Fig. 48) to couple the light into a planar waveguide. If the coupling condition is fulfilled the input beam is spread out into guided and reflected modes (Fig. 46). Application of an electric field to the electrodes leads to the change of waveguide refractive index and consequently to the change of coupling conditions and another redistribution of light intensity between reflected and guided modes. Thus, the intensity of guided and reflected modes depends on the electric field strength. The transmission band of such a modulator depends again on the design of electrodes [196]. The same structure may be used for light deflection [197] as shown in Fig. 49.

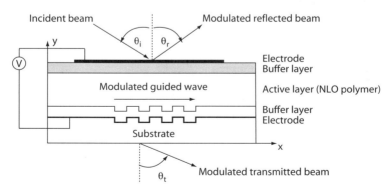

Fig. 48. Schematic representation of an electro-optic modulator using a planar waveguide with corrugated grating

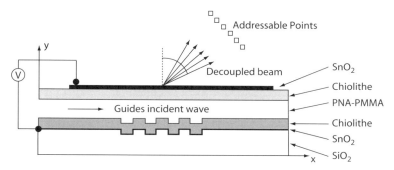

Fig. 49. Schematic representation of a light deflector using a planar waveguide with corrugated grating

8.4
Ultrashort Electric Pulse Generation

Another application of noncentrosymmetric materials is generation of T Hz electric pulses, impossible to obtain by using classical electric circuitry. This can be done either by linear optical rectification effect with $\chi^{(2)}(0;-\omega,\omega)$ susceptibility or through the difference frequency generation (DFG) with $\chi^{(2)}(\omega_1-\omega_2;\omega_1,\omega_2)$ term. In the first case, the femtosecond laser pulse propagating in a nonlinear material creates a DC field of the same temporal duration. In the second case, by using two monochromatic laser pulses close in wavelength, the electromagnetic waves are generated in the submillimeter range. To our knowledge no demonstration was made with poled polymer films but only with highly nonlinear organic crystal, DAST (4'-dimethylamino-N-methyl-4-stilbazolium tosylate) or inorganic crystals (ZnTe) [198, 199].

9
Concluding Remarks

In this review, it was impossible to cite all papers dealing with synthesis of chromophores for SHG, material processing, characterization, and applications, since these are too numerous. Many more applications with these materials have been demonstrated, for example, high-electric field measurement [200]. This field of research is still experiencing rapid development, not only in theoretical design and synthesis of new highly efficient chromophores, better stability, high-glass transition temperature polymers, but also in device applications. The problems encountered with these materials are more clear now than at the beginning of this approach in the early eighties. Although a large progress is noticed in getting materials with a large NLO response and improved temporal stability of the induced polar order, the problem of the chemical (aging) and photochemical stability (light-induced chemical reactions and/or molecule decomposition) requires more systematic studies and a better understanding to be solved [201, 202]. It requires a closer interdisciplinary collaboration between chemists, physicists, and device engineers. The solution to this problem is particularly important for practical applications of these materials.

Acknowledgement. K.-S. Lee acknowledges to the Korea Science Engineering Foundation (KOSEF) through the Center for Advanced Functional Polymers (CAFPoly) for financial support.

References

1. Shi RF, Garito AF (1998) Characterization techniques and tabulations for organic nonlinear optical materials. In: Kuzyk M, Dirk C (eds) Marcel Dekker, New York, p 767
2. Zyss J (1991) Nonlinear Opt 1:3
3. Zyss J (1991) J Chem Phys 98:6583
4. Liptay W (1969) Angew Chem Int Ed 8:177
5. Gonin D, Noël C, Kajzar F (1994) Nonlinear Opt 8:37
6. Oudar JL (1977) J Chem Phys 8:117
7. Singer KD, Garito AF (1981) J Chem Phys 75:3572
8. Kajzar F, Ledoux I, Zyss J (1987) Phys Rev B 36:2210
9. Kajzar F (1997) Electric field induced second harmonic generation; in materials characterization and optical probe techniques. In: Lessard R, Franke H (eds) Critical reviews of optical science and technology. SPIE, vol CR69, Bellingham, p 480
10. Kajzar F, Chollet PA (1997) Poled polymers and their application in second harmonic generation and electro-optic modulation devices, In: Miyata S, Sasabe H (eds) Poled polymers and their application to SHG and EO devices. Gordon & Breach, Amsterdam, p 1
11. Messier J, Kajzar F, Sentein C, Barzoukas M, Zyss J, Blanchard-Desce M, Lehn JM (1992) Nonlinear Opt 2:53
12. Burland D, Walsh A, Kajzar F, Sentein C (1991) J Opt Soc Am B 8:2269
13. Cyvin SJ, Rauch JE, Decius JC (1965) J Chem Phys 63:3285
14. Terhune RW, Maker PD, Savage CM (1970) Phys Rev Lett 14:681
15. Clays C, Persoons A (1991) Phys Rev Lett 66:2980
16. Dantas de Morais T, Noël C, Kajzar F (1996) Nonlinear Opt 15:315
17. Swalen J, Kajzar F (1996) Organic thin films for waveguiding nonlinear optics: science and technology. In: Kajzar F, Swalen J (eds) Gordon & Breach, Amsterdam, p 1
18. Teng CC, Man HT (1990) Appl Phys Lett 56:1734
19. Schildkraut JS (1990) Appl Opt 29:2839
20. Levy Y, Dumont M, Chastaing E, Robnin P, Chollet PA, Gadret G, Kajzar F (1993) Nonlinear Opt 4:1
21. Marder SR, Kippelen B, Jen AK-Y, Peyghambarian N (1997) Nature 388:845
22. Oudar JL, Chemla DS (1977) J Chem Phys 66:2664
23. Marder SR, Cheng LT, Tiemann BG, Friedli AC, Blanchard-Desce M, Perry JW, Skindhøj J (1994) Science 263:511
24. Jen AK-Y, Rao VP, Wong KY, Drost KJ (1993) J Chem Soc Chem Commun 90
25. Ahlheim M, Barzoukas M, Bedworth PV, Hu JY, Marder SR, Perry JW, Stahelin CW, Zysset B (1996) Science 271:335
26. Moylan CR, Ermer S, Lovejoy SM, McComb I-H, Leung DS, Wortmann R, Krdmer P, Twieg RJ (1996) J Am Chem Soc 118:12950
27. Zhang C, Wang CG, Yang JL, Dalton LR, Sun GL, Zhang H, Steier WH (2001) Macromolecules 34:235

28. Shi YQ, Zhang C, Zhang H, Bechtel JH, Dalton LR, Robinson BH, Steier WH (2000) Science 288:119
29. Dalton LR (2000) Opt Eng 39:589
30. Wu JW, Valley JF, Ermer S, Binkley ES, Kenney JT, Lipscomb GF, Lytel R (1991) Appl Phys Lett 58:225
31. Jen AK-Y, Wong KY, Rao VP, Drost K, Cai YM (1994) J Electron Mater 23:653
32. Cai YM, Jen AK-Y (1995) Appl Phys Lett 117:7295
33. Katz HE, Dirk CW, Schelling ML, Singer KD, Sohn CJ (1987) Mater Res Soc Symp Proc 109:127
34. Wu XM, Wu JY, Liu YQ, Jen AK-Y (1999) J Am Chem Soc 121:472
35. Ye C, Marks TJ, Yang J, Wong GK (1987) Macromolecules 20:2322
36. Singer KD, Kuzyk MG, Holland WR, Sohn JE, Lalama SL, Comizzoli RB, Katz HE, Schilling ML (1988) Appl Phys Lett 53:1800
37. Jacobson S, Landi P, Findakly T, Stamatoff J, Yoon H (1994) J Appl Polym Sci 53:649
38. Michelotti F, Toussaere E, Levenson R, Liang J, Zyss J (1996) J Appl Phys 80:1773
39. Choi DH, Kang JS, Hong HT (2001) Polymer 42:793
40. Park KH, Song S, Lee YS, Lee CJ, Kim N (1998) Polym J 30:470
41. Jin S, Wübbenhorst M, Van Turnhout J, Mijs W (1996) Macromol Chem Phys 197:4135
42. Beltrami T, Bösch M, Centore R, Concilio S, Günter P, Sirigu A (2001) Polymer 42:4025
43. Yoon C-B, Jung B-J, Shim H-K (2001) Synth Met 117:233
44. Becker M, Sapochak L, Dalton L, Steier W, Jen A K-Y (1994) Chem Mater 6:104
45. Saadeh H, Gharari A, Yu D, Yu L (1997) Macromolecules 30:5403
46. Jiang H, Kakkar AK (1998) Macromolecules 31:4170
47. Van den Broeck K, Verbiest T, Van Beylen M, Persoons A, Samyn C (1999) Macromol Chem Phys 200:2629
48. Saadeh H, Yu D, Wang LM, Yu LP (1999) J Mater Chem 9:1865
49. Li Z, Zhao Y, Zhon J, Shen Y (2000) Eur Polym Sci 36:2417
50. Davey MH, Lee VY, Wu LM, Moylan CR, Volksen W, Knoesen A, Miller RD, Marks TJ (2000) Chem Mater 12:1679
51. Woo HY, Shim H-K, Lee K-S, Jeong M-Y, Lim T-K (1999) Chem Mater 11:218
52. Leng WN, Zhou YM, Xu QH, Liu JZ (2001) Polymer 42:9253
53. Van den Broeck K. Verbiest T, Degryse J, Van Beylen M, Persoons A, Samyn C (2001) Polymer 42:3315
54. Nemoto N, Miyata F, Nagase Y (1996) Chem Mater 8:1527
55. Nemoto N, Miyata F, Kamiyama T, Nagase Y, Abe J, Shirai Y (1999) Macromol Chem Phys 200:2309
56. Nemoto N, Miyata F, Nagase Y, Abe J, Hasegawa M, Shirai F (1996) Macromolecules 29:2365
57. Nemoto N, Miyata F, Nagase Y, Abe J, Hasegawa M, Shirai F (1996) Mater Chem 6:711
58. Sekkat Z, Kang CS, Aust EF, Wegner G, Knoll W (1995) Chem Mater 7:142
59. Fu CYS, Lackritz HS, Priddy DB, McGrath JE (1996) Chem Mater 8:514
60. Jen AK-Y, Wu X, Ma H (1998) Chem Mater 10:471
61. Ma H, Wang X, Wu X, Liu S, Jen AK-Y (1998) Macromolecules 31:4049
62. Ma H, Jen AK-Y, Wu J, Wu X, Liu S, Shu CF, Dalton LR, Marder SR, Thayumanavan S (1999) Chem Mater 11:2218
63. Kim YW, Lee K-S, Choi KY (1993) Synth Met 57:3998
64. Lee K-S (1997) Macromol Symp 118:519
65. Moon K-J, Shim H-K, Lee K-S, Zieba J, Prasad PN (1996) Macromolecules 29:861
66. Lee S-B, Lee K-S (1999) Private communication of unpublished results
67. Lee S-B, Lee K-S (1999) Nonlinear Opt 22:43
68. Woo HY, Shim H-K, Lee K-S (1998) Macromol Chem Phys 199:1427
69. Woo HY, Shim H-K, Lee K-S (2000) Polym J 32:8
70. Singer KD, Kuzyk MG, Sohn JE (1987) J Opt Soc Am B 4:9680
71. Saleh BEA, Teich MC (1991) Fundamentals of photonics. Wiley, New York, p 780

72. Mader SR, Perry JW, Schaefer WP (1989) Science 245:626
73. Mader SR, Perry JW, Yakymyshyn CP (1994) Chem Mater 6:1137
74. Duan X-M, Konami H, Okada S, Oikawa H, Matsuda H, Nakanishi H (1996) J Phys Chem 100:17780
75. Bosshard Ch (1996) Adv Mater 8:385
76. Yoshima T (1987) J Appl Phys 62:2028
77. Okada S, Masaki A, Matsuda H, Nakanishi H, Kato M, Muranatsu R, Otsuka M (1990) JP J Appl Phys 29:1112
78. Bosshard CH, Knoepfle G, Pretre P, Follonier S, Serbutoviez C, Guenter P (1995) Opt Eng 34:1951
79. Pan F, Wong MS, Bosshard Ch, Guenter P (1986) Adv Mater 8:592
80. Kim O-K, Choi LS, Zhang H-Y, He X-H, Shih Y-H (1966) J Am Chem Soc 118:12,220
81. Clays K, Olbrechts G, Munters T, Persoons A, Kim O-K, Choi LS (1998) Chem Phys Lett 293:337
82. Ashwell, GA, Hargreaves RC, Baldwin CE, Bahra GS, Brown CR (1992) Nature 357:393
83. Kakkar AK, Yitzchaik S, Roscoe SB, Kubota F, Allan DS, Marks TJ, Lin W, Wong GK (1993) Langmuir 9:388
84. Katz HE, Wilson WL, Scheller G (1994) J Am Chem Soc 116:6636
85. Yoon C-B, Moon K-J, Shim H-K, Lee K-S (1998) Mol Cryst Liq Cryst 316:43
86. Horinonchi S, Lee K-S, Lee J-H, Sasaki K (2000) Nonlinear Opt 25:231
87. White DM, Takekoshi T (1981) J Polym Sci Polym Chem Ed 19:1635
88. Maiti S, Mandal BK (1985) Macromol Chem Ripid Comm 6:841
89. White WA, Weissman SR, Kambour RP (1982) J Appl Poly Sci 27:2675
90. Johnson RO, Burlhis HS (1983) J Polym Sci Polym Symp 70:129
91. Lee K-S, Choi K-Y, Won JC, Park BK, Lee I-T (1993) US Patent 5 212 277
92. Lee K-S, Moon K-J, Woo HY, Shim H-K (1997) Adv Mater 9:978
93. Moon K-J, Lee K-S, Shim H-K private communication of unpublished results
94. Chen M, Yu L, Dalton LR, Shi Y, Steier WH (1991) Macromolecules 24:5421
95. Yoon C-B, Moon K-J, Shim H-K, Lee K-S (1998) Mol Cryst Liq Cryst 316:43
96. Kim T-D, Lee K-S, Lee GU, Kim O-K (2000) Polymer 41:5237
97. Kim T-D, Lee K-S, Jeong YH, Jo JH, Chang S (2001) Synth Met 117:307
98. Shi Y, Steier WH, Chen M, Yu L, Dalton LR (1992) Appl Phys Lett 60:2577
99. Liang Z, Dalton LR, Garner SM, Kalluri S, Chen A, Steier WH (1995) Chem Mater 7:941
100. Trollsås M, Orrenius C, Sahlén F, Gedde UW, Norin T, Hult A, Hermann D, Rudquist P, Komitov L, Lagerwall ST, Lindström J (1996) J Am Chem Soc 118:8542
101. Mao SSN, Ra Y, Guo L, Zhang C, Dalton LR, Chen A, Garner S, Steier WH (1998) Chem Mater 10:146
102. Zhang C, Wang C, Dalton LR, Zhang H, Steier WH (2001) Macromolecules 34:253
103. Park C-K, Zieba J, Zhao C-F, Swedek B, Wijekoon WMKP, Prasad PN (1995) Macromolecules 28:3713
104. Lee K-S, Choi S-W, Woo HY, Moon K-J, Shim H-K, Jeong M-Y, Lim T-K (1998) J Opt Soc Am B 15:393
105. Lee K-S, Samoc M, Prasad PN (1992) Polymer for photonics applications. In: Aggarwal SL, Russo S (eds) Comprehensive polymer science. Pergamon Press, Oxford, Chap 16
106. Lee K-S, Moon K-J, Shim H-K (1996) Korea Polym J 4:191
107. Hecht S, Fréchet JMJ (2001) Angew Chem Int Ed 40:74
108. Vögtle F, Gestermann S, Hesse R, Schwierz H, Windisch B (2000) Prog Polym Sci 25:987
109. Bosman AW, Janssen HM, Meijer EW (1999) Chem Rev 99:1665
110. Newkome GR, He E, Moorefield CN (1999) Chem Rev 99:1689
111. Tomalia DA (1994) Adv Mater 6:529
112. Yokoyama S, Nakahama T, Otoma A, Mashiko S (2000) J Am Chem Soc 122:3174
113. Yokoyama S, Nakahama T, Otoma A, Mashiko S (1998) Thin Solid Films 331:248
114. Bozec HL, Bouder TL, Maury O, Bondon A, Ledoux I, Deveau S, Zyss J (2001) Adv Mater 13:1677

115. Moylan CR, Miller RD, Twieg RJ, Ermer S, Lovejoy SM, Leung DS (1995) Proc SPIE 2527:150
116. Harper AW, Sun S, Dalton LR, Garner SM, Chen A, Kalluri S, Steier WH, Robinson BH (1998) J Opt Soc Am B 15:329
117. Robinson BH, Dalton LR (2000) J Phys Chem A 104:4785
118. Ma H, Chen BQ, Sassa T, Dalton LR, Jen AK-Y (2001) J Am Chem Soc 123:986
119. Hsiue G-H, Lee R-H, Jeng R-J (1997) Chem Mater 9:883
120. Yoon C-B, Lee Y-B, Shim H-K, Mun J, Yoon CS (1998) Korea Polym J 6:325
121. Sung PH, Hsu T-F (1998) Polymer 39:1453
122. Jiang H, Kakkar AK (1998) Macromolecules 31:4170
123. Jiang H, Kakkar AK (1998) Macromolecules 31:2501
124. Jiang H, Kakkar AK (1998) Adv Mater 10:1093
125. Jiang H, Kakkar AK (1999) J Am Chem Soc 121:3657
126. Kim HK, Kang S-J, Choi S-K, Min Y-H, Yoon CS (1999) Chem Mater 11:779
127. Xu L, Hou Z, Liu L, Xu Z, Wang W, Li F, Ye M (1999) Opt Lett 24:1364
128. Chaumel F, Jiang H, Kakkar A (2001) Chem Mater 13:3389
129. Kuo W-J, Hsiue G-H, Jeng R-J (2001) Macromol Chem Phys 202:1782
130. Zhang Y, Prasad PN, Burzynski R (1992) Chem Mater 4:851
131. Min YH, Lee K-S, Yoon CS, Do LM (1998) J Mater Chem 8:1225
132. Na SH, Min YH, Yoon CS, Lee J-H, Lee K-S (1998) Appl Phys (Korea) 11:50
133. Min YH, Mun J, Yoon CS, Kim H-K, Lee K-S (1999) Electron Lett 35:1770
134. Lee K-S, Kim T-D, Min YH, Yoon CS, Samoc M, Samoc A (2000) Mol Cryst Liq Cryst 353:525
135. Kim T-D, Lee K-S, Lee SY, Kim YJ, Song JW (2001) Mol Cryst Liq Cryst 371:337
136. Sekkat Z, Dumont M (1992) Nonlinear Opt 2:359
137. Charra F, Kajzar F, Nunzi JM, Raimond P, Idiart E (1993) Opt Lett 18:941
138. Comizzoli B (1987) J Electrochem Soc Solid Sci Technol 134:424
139. Singer KD, Kuzyk MG, Holland WR, Sohn JE, Lalama SJ, Comizzoli B, Katz HE, Schilling ML (1988) Appl Phys Lett 53:1800
140. Date M, Furukawa T, Yamaguchi T, Kojima A, Shibata I (1989) IEEE Trans Dielect El In 24:537
141. Yilmaz S, Bauer S, Gerhard-Multhaupt R (1994) Appl Phys Lett 64:2770
142. Gross, B, Gerhard-Multhaupt R, Berraisoul A, Sessler GM (1987) J Appl Phys 62:1429
143. Bauer S (1996) Appl Phys Rev 80:5531
144. Kajzar F, Nunzi JM (1998) Molecule orientation techniques. In: Kajzar F, Reinisch R (eds) Beam shaping and control with nonlinear optics. Plenum Press, New York, p 101
145. Sprave M, Blum R, Eich M (1996) Appl Phys Lett 69:2962
146. Blum RM, Sablotny J, Eich M (1998) J Opt Soc Am B 15:318
147. Houe M, Townsend PD (1995) J Phys D 28:1747
148. Yang GM, Bauer-Gorgonea S, Sessler GM, Bauer S, Ren W, Wirges W, Gerhard-Multhaupt R (1994) J Appl Phys 64:22
149. Norwood RA, Khanarian G (1990) Electron Lett 26:210
150. Mortazavi MA, Knoesen A, Kowel ST, Higgins BG, Dienes A (1989) J Opt Soc Am B 6:733
151. Gadret G, Kajzar F, Raimond P (1991) Nonlinear optical properties of poled polymers. In: Singer KD (ed) Nonlinear optical properties of organic materials IV. Proc SPIE 1560:226
152. Dantas de Morais T, Noël C, Kajzar F (1996) Nonlinear Opt 15:315
153. Gonin D, Guichard B, Large M, Dantas de Morais T, Noël C, Kajzar F (1996) J Non-linear Opt Phys Mater 5:735
154. Noël C, Kajzar F (1998) Highly orientable liquid crystalline polymers for quadratic nonlinear optics. In: Prasad PN, Mark JE, Kandil SH, Kafafi Z (eds) Proceedings of the fourth international conference on frontiers of polymers and advanced materials, Plenum Press, New York, p 275
155. Dumont M, Sekkat Z, Loucif-Saibi R, Nakatani K, Delaire J (1993) Nonlinear Opt 5:229

156. Dumont M, Froc G, Hosotte S (1995) Nonlinear Opt 9:327
157. Wachtveil J, Nägele T, Puell B, Zinnh W, Krüger M, Rudolph-Böhner S, Osterhelt D, Mo-
     roder L (1997) J Photoch Photobio A 105:283
158. Nägele T, Hoche R, Zinnh W, Wachtveil J (1997) Chem Phys Lett 272:489
159. Kajzar F, Charra F, Nunzi JM, Raimond P, Idiart E, Zagorska M (1995) Third-order non-
     linear optical properties of functionalized polymers. In: Prasad PN (ed) Proceedings of
     international conference on frontier of polymers and advanced materials, Plenum Press,
     New York, p 141
160. Dumont M (1996) Mol Cryst Liq Cryst 282:437
161. Fiorini C, Charra F, Nunzi JM, Raimond P (1994) J Opt Soc Am B 11:2347
162. Nunzi JM, Fiorini C, Charra F, Kajzar F, Raimond P (1995) All-optical poling of poly-
     mers for phase-matched frequency doubling. In: Lindsay GA, Singer KD (eds) Polymers
     for second-order nonlinear optics. ACS symposium series, vol 601, American Chemical
     Society, Washington DC, p 240
163. Osterberg U, Margulis W (1986) Opt Lett 11:516
164. Stolen RH, Tom HWK (1987) Opt Lett 12:585
165. Baranova NB, Zeldovich BY (1987) JETP Lett 45:717
166. Kielich S (1969) IEEE J Quantum Elect 5:562
167. Page H, Jurich MC, Reck B, Sen A, Twieg RJ, Swalen JD, Bjorklund GC, Wilson GC
     (1990) J Opt Soc Am B 7:1239
168. Wu JW (1991) J Opt Soc Am B 8:142
169. Kajzar F, Noël C (1998) Adv Mater Opt Electr 8:247
170. Kohlrausch F (1847) Pogg Ann Physk 91:179
171. Williams G, Watts DC (1970) Trans Faraday Soc 66:80
172. Large M, Kajzar F, Raimond P (1998) Appl Phys Lett 73:3635
173. Combellas C, Mathey G, Petit MA, Thiebault A, Kajzar F (1998) Photonics Sci News 3:5
174. Combellas C, Kajzar F, Mathey G, Petit MA, Thiebault A (2000) Chem Phys 252:165
175. Sasaki K, Zhang GJ, Horinouchi S (1994) Polymer films obtained by DC-electric field
     and pure optical poling. In: Möhlmann GR (ed) Nonlinear optical properties of organic
     materials VII. Proc SPIE, vol 2285, p 290
176. Bosshard C, Küpfer M, Flörsheimer M, Günter P (1991) Opt Commun 85:247
177. Azumai Y, Seo I, Sato H (1992) IEEE J Quantum Elect 28:231
178. Khanarian G, Norwood RA, Landi P (1990) In: Khanarian G (ed) Nonlinear optical
     properties of organic materials II. Proc SPIE, vol 1147, p 129
179. Khanarian G, Norwood RA, Haas D, Feuer B, Karim D (1990) Appl Phys Lett 57:977
180. Otomo A, Mittler-Neher S, Bosshard Ch, Stegeman GI, Horsthuis H, Möhlmann GR
     (1993) Appl Phys Lett 63:3405
181. Möhlmann G, Horsthuis HG, Otomo A, Stegeman GI (1994) Polymer films obtained by
     DC-electric field and pure optical poling. In: Möhlmann GR (ed) Nonlinear optical
     properties of organic materials VII. Proc SPIE, vol 2285, p 300
182. Alshikh Khalil M, Vitrant G, Raimond P, Chollet PA, Kajzar F (1999) Opt Commun
     170:281
183. Normandin R, Stegeman GI (1979) Opt Lett 4:58
184. Normandin R, Stegeman GI (1980) Appl Phys Lett 36:253
185. Normandin R, Stegeman GI (1982) Appl Phys Lett 40:759
186. Normandin R, Letorneau S, Chatenoud F, Williams RL (1991) J Quantum Elect 27:1250
187. Blau G, Popov E, Kajzar F, Raimond P, Roux JF, Coutaz JL (1995) Opt Lett 20:1101
188. Dalton LR, Harper AW (1996) Photoactive organic materials for electro-optic modulator
     and high density optical memory applications. In: Kajzar F, Agranovich VM, Lee CYC
     (eds) Photoactive organic materials: science and application. NATO ASI Series, vol 9,
     Kluwer Academic, Dordrecht, p 183
189. Ashley PR (1996) Component integration and applications with organic polymers. In:
     Kajzar F, Agranovich VM, Lee CYC (eds) Photoactive organic materials: Science and ap-
     plication. NATO ASI Series, vol 9, Kluwer Academic, Dordrecht, p 199

190. Flipse MC, Van der Vorst CPJM, Hofstraat JW, Wounderberg RH, Van Gassel RAP, Lamers JC, Van der Linden EGM, Veenis WJ, Diemeer MBJ, Donckers MCJM (1996) Recent progress in polymer based electro-optic modulators: materials and technology. In: Kajzar F, Agranovich VM, Lee CYC (eds) Photoactive organic materials: science and application. NATO ASI Series, vol 9, Kluwer Academic, Dordrecht, p 229

191. Kenney J, Jen AK-Y (1996) Polymer material systems for E/O waveguide devices. In: Kajzar F, Agranovich VM, Lee CYC (eds) Photoactive organic materials: science and application. NATO ASI Series, vol 9, Kluwer Academic, Dordrecht, p 209

192. Ferm P, Knapp CW, Wu C, Yardley JT, Hu BB, Zhang X, Austin DH (1991) Appl Phys Lett 59:1159

193. Dalton LR, Harper AW, Wu B, Ghosn R, Laquindanum J, Liang Z, Hubbel A, Xu C (1995) Adv Mat 7:519

194. Dalton LR (2002) Adv Polym Sci 158:1

195. Blau G, Vitrant G, Chollet PA, Raimond P, Kajzar F (1997) Functionalized polymers for electro-optic modulation through grating induced resonant excitation of guided modes. In: Lampropoulos GA, Lessard RA (eds) Applications of photonic technology 2: Communications, sensing, materials and signal processing. Plenum Press, New York, p 15

196. Blau G, Cairone L, Ruiz L, Vitrant G, Chollet PA, Kajzar F (1997) Electro-optic modulation through grating induced resonant excitation of guided modes. In: Möhlmann G (ed) Nonlinear optical properties of organic molecules IX. Proc SPIE, vol 2852, p 237

197. Blau G, Kajzar F, Raimond P, Vitrant G (1997) Deflecteur electro-optique de faisceaux lumineux, notamment pour adressage optique multipoints. French Patent 97 07045

198. Zhang XC, Ma XF, Jin Y, Lu TM, Boden EP, Phelps DP, Stewart KR, Yakymyshin CP (1992) Appl Phys Lett 61:3080

199. Kawase K, Mizuno M, Sohma S, Takahashi H, Taniuchi T, Urata Y, Wada S, Tashiro H, Ito H (1999) Opt Lett 24:1065

200. Skindhoj J, Perry JW, Marder SR (1994) Polymer films obtained by DC-electric field and pure optical poling. In: Möhlmann GR (ed) Nonlinear optical properties of organic materials VII. Proc SPIE, vol 2285, p 116

201. Lee CYC (1996) Optimizing secondary properties and EO coefrficients of EO polymers, In: Kajzar F, Agranovich VM, Lee CYC (eds) Photoactive organic materials: Science and application. NATO ASI Series, vol 9, Kluwer Academic, Dordrecht, p 175

202. Stegeman GI, Galvan-Gonzales A, Canva M, Twieg R, Kowalczyk AC, Zhang XQ, Lackritz HS, Marder S, Thayumanavan S, Chang KP, Jen AK-Y, Wu X (2000) Nonlinear Opt 25:57

Received April 2002

Photorefractive Polymers and their Applications

Bernard Kippelen* · Nasser Peyghambarian

Optical Sciences Center, The University of Arizona, 1630 E. University Blvd, Tucson AZ 85721, USA
*E-mail:kippelen@u.arizona.edu

Photorefractive polymers exhibit large refractive index changes when exposed to low power laser beams. When the optical excitation consists of two interfering coherent beams, the periodic light distribution produces a periodic refractive index modulation. The resulting index change produces a hologram in the volume of the polymer film. The hologram can be reconstructed by diffracting a third laser beam on the periodic index modulation. In contrast to many physical processes that can be used to generate a refractive index change, the photorefractive effect is fully reversible, meaning that the recorded holograms can be erased with a spatially uniform light beam. This reversibility makes photorefractive polymers suitable for real-time holographic applications. The mechanism that leads to the formation of a photorefractive index modulation involves the formation of an internal electric field through the absorption of light, the generation of carriers, their transport and trapping over macroscopic distances. The resulting electric field produces a refractive index change through orientational or non-linear optical effects. Due to the transport process, the index modulation amplitude is phase shifted with respect to the periodic light distribution produced by the interfering optical beams that generate the hologram. This phase shift enables the coherent energy transfer between two beams propagating in a thick photorefractive material. This property, referred to as two-beam coupling, is used to build optical amplifiers. Hence, photorefractive materials are also playing a role in imaging applications. Discovered and studied for several decades mainly in inorganic crystals and semiconductors, the photorefractive effect has not yet found wide spread commercial applications. This can be attributed to the difficulties associated with the growth of crystals, and to the high cost of optical and optomechanical components necessary for the development of complete optical systems. With the emergence of novel low cost plastic optical components that can be mass produced by techniques such as injection molding, the cost and the weight of optical components is decreasing rapidly. This trend together with the advances made in fabricating integrated laser sources at lower cost provide a great momentum to the development of new optical processing technologies. As real-time optical recording and processing media, photorefractive polymers are expected to play a major role in these technologies. The optical, physical, and chemical properties of photorefractive polymers are outlined and discussed. Current material classes and their respective merits and future challenges are presented together with examples of applications.

Keywords. Polymer, Photorefractive, Photoconductive, Charge generation, Carrier transport, Electro-optics, Orientational birefringence, Kerr effect, Real-time holography, Interferometry, Holographic storage, Glass transition temperature, Optical processing, Four-wave mixing, Two-beam coupling.

List of Abbreviations and Symbols 89

1   **Introduction.** 92

2   **Linear Optical Properties of Polymers** 93

2.1   Electromagnetic Theory of Light. 93
2.1.1   Maxwell's Equations and the Constitutive Equations 93
2.1.2   Optics of Anisotropic Media 98

3   **Introduction to Nonlinear Optics and Electro-Optics** 100

3.1   Fundamentals of the Nonlinear Optics Formalism. 100
3.2   Fundamentals of Electro-Optics 105

4   **Molecular and Bulk Nonlinear Optics** 108

4.1   Microscopic Theory: First and Second Hyperpolarizability 108
4.2   Two-Level Model 109
4.3   Bond-Length Alternation Model. 112
4.4   Oriented Gas Model. 114
4.4.1   Electro-Optic Properties 114
4.4.2   Orientational Birefringence. 118

5   **Photorefractive Effect** 120

5.1   The Kukhtarev Model. 124
5.2   Four-Wave Mixing and Two-Beam Coupling 128

6   **Organic Photorefractive Materials** 131

6.1   Design of Photorefractive Polymers 133
6.2   Physical and Chemical Properties of Photorefractive Polymers 135
6.2.1   Photogeneration. 135
6.2.2   Transport. 136
6.2.3   Orientational Photorefractivity 137
6.3   Examples of Organic Photorefractive Materials. 140

7   **Applications**. 148

8   **Conclusion and Outlook**. 152

References 153

List of Abbreviations and Symbols

E	electric field
H	magnetic field
D	electric displacement field
B	magnetic induction field
c	speed of light
ϱ	electric charge density
j	current density
$\nabla \cdot$	divergence operator
$\nabla \times$	curl operator
ε	dielectric function
P	polarization
μ	magnetic permeability
χ	optical susceptibility
n'	complex refractive index
n	real part of complex refractive index
k'	imaginary part of complex refractive index
$U(r, t)$	complex wavefunction
v	speed of light in a medium with refractive index n
ω, ω_i	optical frequency
k	wavevector
λ	wavelength
ζ, E	electric field complex amplitude
α	absorption coefficient
e	elementary charge
m	electron rest mass, and fringe visibility
Γ'	damping constant
ω_0	resonance frequency
N	density of electrons, density of molecules
χ'	real part of the complex susceptibility
χ''	imaginary part of the complex susceptibility
P	Cauchy principal value
$\tilde{I}$	unitary second-rank tensor
$\tilde{\varepsilon}$	dielectric tensor
n_X, n_Y, n_Z	principal indices of refraction
n_o	ordinary refractive index
n_e	extraordinary refractive index
θ	angle
χ_L	linear part of the total optical susceptibility
$\chi^{(2)}$	second-order nonlinear susceptibility
$\chi^{(3)}$	third-order nonlinear susceptibility

D	degeneracy factor
d_{ij}	second-order nonlinear tensor elements
E_0	d.c. or low frequency electric field
r, r_{ij}	linear electro-optic coefficient, electro-optic tensor elements
s	quadratic electro-optic coefficient and photoionization cross-section
$\tilde{\eta}, \eta_{ij}$	impermeability tensor, impermeability tensor elements
μ, μ_i	dipole moment, dipole moment component
$\tilde{\alpha}, \alpha_{ij}$	linear molecular polarizability tensor, tensor components
$\tilde{\beta}, \beta_{ijk}, \beta_{zzz}, \beta$	first hyperpolarizability tensor, tensor components
$\tilde{\gamma}$	second hyperpolarizability tensor
$R^{(n)}$	degeneracy factors
β_{add}, β_{CT}	additive and charge-transfer contributions to the first hyperpolarizability
f	optical transition oscillator strength and lens focal length
$\Delta\mu_{ge}$	difference in dipole moment between excited state and ground state
$\hbar\omega_{ge}$	photon energy
μ_{ge}	transition dipole moment
$\beta(0)$	dispersion-free value of the first hyperpolarizability
$\hbar = h/2\pi$	Planck constant
E_{ge}	energy
$\lvert\psi>, \lvert n>, \lvert z>$	wave functions
p_i	molecular polarizability component
$\{X,Y,Z\}$	laboratory Cartesian system of coordinates
$\{x,y,z\}$	molecular Cartesian system of coordinates
V	volume
a_{Ii}	director cosines
φ, θ, ψ	Euler angles
$d\Omega$	differential solid angle
$U(\theta)$	interaction energy
$G(\varphi, \theta, \psi), G(\Omega)$	Maxwell-Boltzmann distribution function
$\mu^\star$	corrected value of dipole moment
E_p	poling field
k_B	Boltzmann constant
T	temperature
$L_n(x)$	Langevin functions
$I(x), I_i, I_0$	optical intensity

Λ	grating spacing
e_i	unitary polarization vectors
N_D^0, N_D, N_D^+	density of donors, density of neutral donors, and density of donor radical cations
N_A, N_A^-	density of acceptors, density of acceptor radical anions
β'	thermal generation rate
γ	recombination coefficient
n_e	density of electrons
$\vec{\nabla}$	gradient operator
μ_e	electron drift mobility
ε_{dc}	d.c. or low frequency dielectric constant
$E^{(dc)}, E_1, E_{sc}, E_0, E_m, E_{ext}$	d.c. or low frequency electric field
E_D	diffusion field
E_q	limiting space-charge field
N_{eff}	effective photorefractive trap density
Θ	shift of the photorefractive space-charge field
$\Delta n, \Delta \tilde{n}$	refractive index modulation amplitude, complex refractive index modulation amplitude
r_{eff}	effective electro-optic coefficient
Q	factor for evaluating thick or thin gratings
d	sample thickness
c_i, c_d	obliquity factors
η	diffraction efficiency
α_1, α_2	propagation angles
Γ	two-beam coupling gain coefficient
γ_0	intensity ratio in two-beam coupling
Q'	figure of merit
$\mu(E, T)$	carrier mobility
K	grating vector
$\sigma, \sigma_{VdW}, \sigma_D$	energetical disorder
Σ	positional disorder
T_g	glass transition temperature
T_c	dispersive to nondispersive transition temperature
$\Delta \alpha$	polarizability anisotropy
Q_{OP}	figure of merit for orientational photorefractivity
N_{max}	maximum non-interacting chromophore density
ξ	angle between the total field and the Z axis

1
Introduction

With regard to the pervasiveness of optical technologies, the high bandwidth of optical fibers has enabled the successful development of the internet that is changing the world's economy and the way people communicate and interact. As optics and lasers have invaded our homes and are generating multitrillion-dollar worldwide markets, the leading role of optics is that of a technological enabler because it usually plays a supporting role in the conception of larger systems [1]. But what is the major enabler for optics? Materials. And what kind of new materials have heavily impacted society during the past 30 years? Plastics or synthetic polymers. So, polymers are expected to play a key role in the development of future optical technologies. Optical polymers will enable the emergence of entirely new products in a variety of areas ranging from information technology and telecommunications, health care, to optical sensing and lighting.

During the past decade, polymers for photonic devices have matured considerably and numerous functionalities such as electrical, electro-optical, and light-emitting properties have been incorporated successfully into polymers through the intimate collaboration between scientists from optical sciences, physics, chemistry, and materials science. Photorefractive polymers belong to these new classes of materials with several functionalities that have emerged during the last decade. Capitalizing on the previous development of electro-optic and photoconducting polymers for optical switches, and xerography, respectively, photorefractive polymers did experience a rapid development. Within ten years, they reached a performance level that outperforms some of the best inorganic materials. Active multidisciplinary research efforts in this field are constantly providing a better understanding of the basic principles of photorefractivity in these complex materials.

In this chapter we review the basic optical properties required to understand photorefractivity in polymers. Section two describes the electromagnetic theory of light and introduce basic concepts related to material properties such as *dielectric constant, optical susceptibility, refractive index, absorption coefficient*, and how they relate to each other. Expressions for these quantities are derived using a simple oscillator model in which electrons bound to positively charged ions are displaced by the oscillating electric field of a light beam. This simple model provides a qualitative description of many key aspects of light-matter interaction and yields realistic predictions for systems that are transparent and weakly absorbing like polymers. We also introduce the basic concepts of crystal optics and discuss the linear optical properties of anisotropic media. Tensorial properties of the material properties are introduced as well as crystal symmetry considerations. The third and fourth sections provide a description of the nonlinear optical and electro-optic properties of polymers. These concepts are necessary for the understanding of photorefractivity that is described in section five. While polymers and crystals have very different properties, the *Kukhtarev model* developed initially to describe photorefractivity in crystals provides a good framework for polymers too, and most of the experiments performed in polymers could be fairly well described within this framework. In section six, we describe the specificity of photorefractive polymers and illustrate some of the properties through selected examples of composites that have been developed in recent years. We do not give an exhaustive survey of the different materials reported in recent years as such surveys can be found in several recent review articles [2–4]. The last section will describe some examples of applications that have been demonstrated so far with photorefractive polymers.

2
Linear Optical Properties of Polymers

2.1
Electromagnetic Theory of Light

2.1.1
Maxwell's Equations and the Constitutive Equations

Light beams are represented by electromagnetic waves that are described in a medium by four vector fields: the *electric field* $E(r, t)$, the *magnetic field* $H(r, t)$, the *electric displacement field* $D(r,t)$, and $B(r,t)$ the *magnetic induction field* (or magnetic flux density). Throughout this chapter we will use bold symbols to denote vector quantities. All field vectors are functions of position and time. In a dielectric medium they satisfy a set of coupled partial differential equations known as Maxwell's equations. In the CGS system of units, they give

$$\nabla \times H = \frac{1}{c}\frac{\partial D}{\partial t} + \frac{4\pi}{c}j$$

$$\nabla \times E = -\frac{1}{c}\frac{\partial B}{\partial t}$$

$$\nabla \cdot D = 4\pi\varrho \qquad\qquad\qquad (1)$$

$$\nabla \cdot B = 0$$

where c is the speed of light, ϱ and j are the *electric charge density* (density of free conduction carriers) and the *current density* vector in the medium, respectively. $\nabla\cdot$ and $\nabla\times$ are the divergence and curl operations. Since polymers are non-conducting dielectrics, one assumes that $\varrho = j = 0$. In a non-magnetic medium, the four fields E, D, H, and B are related through the so-called constitutive equations:

$$D = E + 4\pi P = \varepsilon E$$

$$B = \mu H \qquad\qquad\qquad (2)$$

where the constitutive parameter ε is the *electric permittivity* (also called *dielectric function*) and $P = P(r, t)$ is the electric polarization density vector or *polarization* of the medium. For a non-magnetic medium $\mu = 1$. The concept of polarization occupies a central role in optics since it characterizes the response of a medium to the electrical field associated with a light beam that propagates in this medium. In a dielectric medium the electric field displaces electrons from their equilibrium position. Since the medium is neutral, this perturbation creates locally a dipole moment whenever the negative and positive charge densities do not coincide. The polarization of the medium is the macroscopic sum of these electric dipole moments induced by the electric field per unit volume. In organic molecules and polymers the nature of the bonds determine the magnitude of the binding energy and consequently the ability of the electrons involved in these bonds to be distorted by an electric field. Thus, one expects the polarization to be much higher in conjugated materials in which delocalized π electrons are easily polarized compared with saturated materials where σ electrons are more difficult to perturb. On a molecular level, one understands why anisotropic molecules will have strong anisotropic optical properties: a conjugated molecule with a rod-like shape, for instance, will be much easier to polarize with an electric field applied along the molecular axis than when the field is applied in a perpendicular direction. Next we derive additional concepts in dielectrics such as the optical susceptibility and refractive index. Let us first derive these quantities in idealized materials. Such ideal materials are assumed linear, nondispersive, homogeneous, isotropic, and spatially nondispersive.

A dielectric medium is said to be *linear* if the vector field $P(r, t)$ is linearly related to the vector field $E(r, t)$. This approximation is always used in the field of linear optics but fails in the case of nonlinear optics as will be discussed in more details in Sect. 3. A medium is said to be *nondispersive* if its response is instantaneous, meaning that the polarization at time t depends only on the electric field at that same time t and not on prior values of E. In most dielectrics the response time is very short (femtosecond or picosecond response times), but the fact that it is nonzero has huge consequences as will be discussed later. A medium is said to be *homogeneous* if the response of the material to an electric field is independent of r. A medium is said to be *isotropic* if the relation between E and P is independent of the direction of the field vector E. In the simplest case, when the medium satisfies all these conditions, the vectors P and E at any position and at any time are parallel and proportional and related to each other by

$$P(r,t) = \chi E(r, t) \tag{3}$$

where χ is a scalar and is called the *optical susceptibility*. Substituting Eq. (3) into Eq. (2) gives the following relationship between the susceptibility and the dielectric function:

$$\varepsilon = 1 + 4\pi\chi \tag{4}$$

Note that under the conditions fixed at the beginning of this sub-section, the vectors E, D, and P are all parallel and proportional. Later we will see that this is not the case when the material is no longer isotropic. With Maxwell's and the constitutive equations, one can derive for an isotropic medium the following wave equation:

$$\nabla^2 E - \frac{n^2}{c^2}\frac{\partial^2 E}{\partial t^2} = 0 \tag{5}$$

with

$$n = \sqrt{1 + 4\pi\chi} = \sqrt{\varepsilon} \tag{6}$$

where n represents the *refractive index* of the medium. Note the universality of Eq. (5): it is independent of the system of units and has the general form of a wave equation:

$$\nabla^2 U(r,t) - \frac{1}{v^2}\frac{\partial^2 U(r,t)}{\partial t^2} = 0 \tag{7}$$

where $U(r, t)$ is the complex wavefunction with an amplitude and a phase. In Eq. (7), v is the speed of propagation of the wave. Comparing Eqs. (6) and (7) one sees that the optical wave in a dielectric medium with refractive index n propagates at a speed of $v = c/n$. Since the speed of light is an upper limit, the refractive index of a dielectric material is necessarily higher than unity. To satisfy the wave equation, the electric field vector components must be harmonic functions of time with frequency ω and harmonic functions of space. For a plane wave, the electric field vector gives

$$E(r,t) = \varsigma\cos(kr - \omega t) = \text{Re}\left\{\varsigma e^{i(kr-\omega t)}\right\} = \frac{1}{2}\left\{\varsigma e^{i(kr-\omega t)} + c.c.\right\} = \left\{E e^{i(kr-\omega t)} + c.c.\right\} \tag{8}$$

where k is the *wavevector* with components (k_X, k_Y, k_Z) and *c.c.* denotes complex conjugate. To satisfy the wave equation, the amplitude of the wavevector referred to as the wavenumber k must satisfy the so-called dispersion relation:

$$k = \frac{\omega}{v} = \frac{n\omega}{c} \tag{9}$$

Note that $k = 2\pi/\lambda$ where λ is the wavelength of the light beam. With the electric field representation conventions adopted in Eq. (8), the intensity I of the light beam is given by

$$I = \frac{nc}{8\pi}|\varsigma|^2 = \frac{nc}{2\pi}|E|^2 \tag{10}$$

In the CGS system of units adopted through this chapter all the field quantities E, D, H, and B have the same units and are expressed in statvolt/cm. Intensities are given in erg/cm^2·s. Conversion into MKS units can be obtained easily by remembering that 1 statvolt = 300 V.

In real materials, for which there is a delay in the response, the polarization induced by an electric field is no longer given by Eq. (3) but is given by the convolution product:

$$P(t) = \int_{-\infty}^{t} \chi(t - \tau)E(\tau)d\tau \tag{11}$$

Using the property that the Fourier transform of a convolution product is equal to the product of the Fourier transforms, one can derive a simple relationship between the polarization and electric field in the Fourier domain:

$$P(\omega) = \chi(\omega)E(\omega) \tag{12}$$

where all the quantities are complex and are functions of the frequency ω of the optical field. Thus, a material that does not respond instantaneously to an electric field is said to be dispersive because all the material properties such as *dielectric constant*, *optical susceptibility*, and *refractive index* are no longer real and constant quantities, but complex and frequency dependent. The real part n of the new complex refractive index $n' = n + ik'$ relates to the conventional index concept developed for ideal non-dispersive materials and the imaginary part k' of the index is related to the absorption coefficient through:

$$\alpha = \frac{2k'\omega}{c} \tag{13}$$

where α is the linear absorption coefficient expressed in cm^{-1}. The frequency dependence of the complex refractive index can be obtained by considering the simple Lorentz oscillator model in which electrons bound to positively charged ions are displaced by the oscillating electric field of a light beam. This simple model provides a qualitative description of many key aspects of light-matter interaction and yields realistic predictions for systems that are transparent and weakly absorbing such as polymers. For an oscillator with resonance frequency ω_0 the real and imaginary parts of the index are given by

$$n = 1 - \frac{Ne^2}{m} \frac{2\pi(\omega^2 - \omega_0^2)}{(\omega^2 - \omega_0^2)^2 + 4\Gamma'^2\omega^2} \tag{14}$$

$$k' = \frac{2\pi Ne^2}{m} \frac{2\Gamma'\omega}{(\omega^2 - \omega_0^2)^2 + 4\Gamma'^2\omega^2} \tag{15}$$

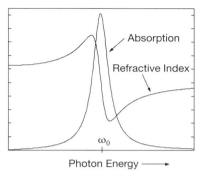

Fig. 1. Real and imaginary parts of the refractive index obtained with the simple oscillator model

where e is the elementary charge, m the mass of an electron, and Γ' is a damping constant. The frequency dependence of the real and imaginary parts of the index are illustrated in Fig. 1.

Note that this simple model predicts the normal dispersion for most optical materials: a decrease of the refractive index (real part) on the low energy side of the resonance (absorption band of a polymer for instance) when the frequency of the light is decreased. Due to causality which is the property that the polymer will not respond to the electric field before the field is applied at a given time t, the real part $\chi'(\omega)$ and imaginary part $\chi''(\omega)$ of the optical susceptibility are forming a Hilbert pair and are therefore related to each other through the following Kramers-Kronig relations [5]:

$$\chi'(\omega) = \frac{2}{\pi} P \int_0^\infty \frac{\omega' \chi''(\omega')}{\omega'^2 - \omega^2} d\omega' \qquad (16)$$

$$\chi''(\omega) = \frac{-2\omega}{\pi} P \int_0^\infty \frac{\chi'(\omega')}{\omega'^2 - \omega^2} d\omega' \qquad (17)$$

where P denotes the Cauchy principal value. These relationships show that if the full frequency dependence of either the real or imaginary parts of the optical susceptibility is known, the other part can be calculated. More importantly, this property shows that absorption changes are always accompanied by refractive index changes and vice versa.

2.1.2
Optics of Anisotropic Media

In the preceding section we have assumed that the material was isotropic, resulting in an induced polarization that is parallel to the electric field. Since most of the polymers are amorphous structures their optical properties are usually isotropic. However, electro-active polymers such as electro-optic and photorefractive polymers are made anisotropic through the orientation of the highly anisotropic functional moieties they contain. These moieties can be incorporated into polymers through different synthetic approaches: (i) the guest/host approach is which functional molecules are doped into the polymer; (ii) the fully functionalized approach where moieties are incorporated into the polymer as side chain pendant groups or into the main chain of the polymer. These functional moieties are noncentrosymmetric and have a permanent dipole moment. Their orientation in an electric field breaks the centrosymmetry in the material and generates simultaneously birefringence and second-order nonlinear optical properties as will be discussed in more details in Sects. 3 and 4. As a result of birefringence, light with different polarization and propagating in different directions will propagate with a different velocity (i.e., experience a different refractive index). Hence, the optical properties of electro-active polymers are those of anisotropic media.

In anisotropic media the polarization is no longer necessarily parallel to the electric field. Polarization and field components are no longer related by scalar proportionality factors but by tensor quantities. In other words, the polarization component in one direction is related to the field components in all three directions. In this case, the relationship between field and polarization for a dispersive medium is given by

$$
\begin{aligned}
P_X &= \chi_{11}E_X + \chi_{12}E_Y + \chi_{13}E_Z \\
P_Y &= \chi_{21}E_X + \chi_{22}E_Y + \chi_{23}E_Z \\
P_Z &= \chi_{31}E_X + \chi_{32}E_Y + \chi_{33}E_Z
\end{aligned}
\quad \Leftrightarrow \quad P_i = \sum_j \chi_{ij} E_j
\tag{18}
$$

where X, Y, and Z represent the three Cartesian coordinates. While the optical susceptibility is described by a scalar in isotropic media (which is a complex number that is frequency dependent if the material is dispersive), in anisotropic media the susceptibility is characterized by nine tensor elements χ_{ij}. The susceptibility in this case is a second-rank tensor with 3^2 elements given by

$$
\tilde{\chi} = \begin{pmatrix}
\chi_{11} & \chi_{12} & \chi_{13} \\
\chi_{21} & \chi_{22} & \chi_{23} \\
\chi_{31} & \chi_{32} & \chi_{33}
\end{pmatrix}
\tag{19}
$$

and the second constitutive relation in an anisotropic medium gives

$$D = \left(\tilde{I} + 4\pi\tilde{\chi}\right)E = \tilde{\varepsilon}E \tag{20}$$

where $\tilde{I}$ is the unitary second-rank tensor. If the light-matter interaction is non-resonant (i.e., the material is transparent and has no absorption) the susceptibility and dielectric tensors are real and symmetric, i.e., $\chi_{ij} = \chi_{ji}$, and each tensor has only six independent elements [6]. The magnitude of these tensor elements depends on the choice of the X, Y, and Z axis relative to the symmetry axis of the anisotropic medium. In this case, it is always possible to find a set of orthogonal axes for which the dielectric tensor is diagonal:

$$\tilde{\varepsilon} = \begin{pmatrix} \varepsilon_{11} & 0 & 0 \\ 0 & \varepsilon_{22} & 0 \\ 0 & 0 & \varepsilon_{33} \end{pmatrix} = \begin{pmatrix} n_X^2 & 0 & 0 \\ 0 & n_Y^2 & 0 \\ 0 & 0 & n_Z^2 \end{pmatrix} \tag{21}$$

where n_X, n_Y, and n_Z are the principal indices of refraction. In the case when all three principal indices are all different, there are two optical axes and the medium is said to be biaxial. Most of the electro-active and photorefractive polymers have only one optical axis which is generally defined by the direction of the field that is used to orient the functional moieties and cylindrical symmetry is assumed around this axis. The symmetry axis is chosen to be Z. Thus, the polymers we are considering in this chapter are uniaxial and are characterized by two principal indices of refraction: an *ordinary* index $n_o = n_X = n_Y$ and an *extraordinary* index $n_e = n_Z$. A symmetric second-rank tensor such as the dielectric tensor (see Eq. 21) can be represented by a quadratic surface that is invariant to the choice of the Cartesian axes and that can be written in the principal axes as

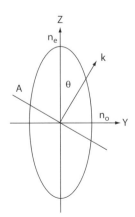

Fig. 2. The index ellipsoid

$$\frac{X^2}{n_X^2} + \frac{Y^2}{n_Y^2} + \frac{Z^2}{n_Z^2} = 1 \tag{22}$$

Equation (22) describes an ellipsoid (see Fig. 2) called the *index ellipsoid*. The latter is very useful in deriving the refractive index of optical waves with different polarization and propagation direction. A wave traveling in a uniaxial polymer at an angle θ with respect to the optic axis experiences two different index depending on its polarization: if the wave is s-polarized (perpendicular to the plane of incidence) the refractive index is n_o and is independent of θ; for a p-polarized wave (polarization in the plane of incidence) the refractive index is given by

$$\frac{1}{n^2(\theta)} = \frac{\cos^2 \theta}{n_o^2} + \frac{\sin^2 \theta}{n_e^2} \tag{23}$$

3
Introduction to Nonlinear Optics and Electro-Optics

3.1
Fundamentals of the Nonlinear Optics Formalism

The field of nonlinear optics started shortly after the discovery of the laser in the early 1960s. The first nonlinear optical experiment was the observation of second-harmonic generation (SHG) by Franken et al. [7]. In this experiment a ruby laser beam was sent onto a quartz crystal and a weak light beam with twice the frequency of the incoming beam was generated. Since then, interest in this field has grown continuously and covers today a variety of processes ranging from fundamental ones, including laser cooling and quantum optics, to more applied ones related to optical information technologies. A broad overview of nonlinear optical processes is beyond the scope of this chapter and we refer the reader to several textbooks that have been published previously on the subject [8–10]. In the following section we will introduce the major concepts and definitions needed to understand photorefractive polymers.

So far, in the description of the interaction of light with matter, we have assumed that the response of the material to an applied optical field was independent of its magnitude. This approximation is valid when the electric field amplitude is negligible compared with the internal electric fields in atoms and molecules. However, when lasers are used as light sources, the intensity of the optical field is usually strong and can drive the electronic response of a dielectric into a nonlinear regime. This nonlinear optical response is described by a field-dependent susceptibility that can be written as

$$\chi(E) = \chi_L + \chi^{(2)}E + \chi^{(3)}EE + \dots \tag{24}$$

where χ_L is the linear part of the total optical susceptibility as defined in Sect. 2. $\chi^{(2)}$ and $\chi^{(3)}$ are the second- and third-order nonlinear optical susceptibilities, respectively. Consequently, the polarization $P(r, t)$ no longer linearly relates to electric field $E(r, t)$ as in Eq. (3) but contains additional contributions that are quadratic and cubic in electric field:

$$P(E) = \chi_L E + \chi^{(2)} EE + \chi^{(3)} EEE + \dots \tag{25}$$

The second and third terms of the right hand side of Eq. (25) constitute the second- and third-order nonlinear contributions to the total polarization. These corrections to the polarization are responsible for numerous nonlinear optical processes such as the generation of light beams with new frequencies or an intensity dependent refractive index.

Nonlinear optical processes are usually classified in two classes: second-order and third-order processes. While all the materials exhibit third-order nonlinear optical properties, symmetric requirements limit the number of second-order materials: only materials that do not possess an inversion symmetry (also called noncentrosymmetry) can exhibit second-order nonlinear optical properties. This condition can be easily understood by considering that in a media with an inversion symmetry (or centrosymmetry) the polarization induced by two electric fields pointing in opposite directions are equal in amplitude but have opposite sign ($P(-E) = -P(E)$). By writing this condition using the definition of the total polarization given by Eq. (25) one can easily demonstrate that $\chi^{(2)} = 0$ in this case. Noncentrosymmetry is a necessary condition for electro-optic and photorefractive materials which are based on second-order nonlinear optical processes. Note the difference between noncentrosymmetry and anisotropy. As shown in Fig. 3, a medium that is optically anisotropic can still be centrosymmetric. Let's consider a dielectric doped with rod-like molecules that have cylindrical symmetry along their axes but with a head different from the tail (like molecules that have a dipole moment pointing in one direction). If the molecules are aligned along a preferred direction, the medium will be optically anisotropic because the polarizability of the molecule along its axis is different from that in a perpendicular direction. However, since the molecules are only aligned, on average half of the molecules have their head pointing in one direction and the medium is invariant under a 180° rotation (this is the case in nematic liquid crystals, for instance). Such a medium will be anisotropic but still centrosymmetric and will not exhibit any second-order nonlinear optical properties. To break the centrosymmetry, orientation of the molecules is required. As opposed to alignment, in the case of orientation all the molecules have their head pointing in one direction as shown in Fig. 3c. Thus, noncentrosymmetric materials are necessarily anisotropic but in contrast anisotropy by itself does not always guarantee second-order nonlinear optical properties.

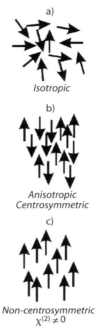

Fig. 3a–c. Schematics of: **a** an isotropic distribution of rod-like molecules; **b** an anisotropic but centrosymmetric distribution; **c** a non-centrosymmetric distribution

Since second-order nonlinear optical materials are anisotropic, their optical properties are described by tensors as discussed previously in Sect. 2.1.2. For a nonlinear optical process, the n-th order nonlinear polarization is due to n interacting electric field vectors and is described by an $(n + 1)$ rank tensor composed of 3^{n+1} tensor elements. In nonlinear optics, several fields with different frequencies ω_n can be present simultaneously so that the electric field and the polarization are represented by

$$E(r,t) = \sum_n E(\omega_n)e^{-i\omega_n t} + c.c \tag{26}$$

$$P(r,t) = \sum_n P(\omega_n)e^{-i\omega_n t} + c.c \tag{27}$$

where the summation is restricted to positive frequencies only and $c.c.$ denotes the complex conjugate. Note that in this representation $E(\omega_n)$ and $P(\omega_n)$ are complex quantities. Note that in the representation given by Eqs. (26) and (27) both electric field and polarization are given by the sum of a complex number and its complex conjugate. This sum is a real number. As for linear dispersive materials (see

Eq. 12), Eq. (25) is valid in the Fourier domain and relates the components of the susceptibility tensor elements and the complex components of the electric field and polarization. All quantities are frequency dependent. The general expression of the second-order nonlinear polarization is given by [10]:

$$P_i(\omega_n + \omega_m) = D \sum_{jk} \chi_{ijk}^{(2)}(\omega_n + \omega_m; \omega_n, \omega_m) E_j(\omega_n) E_k(\omega_m) \tag{28}$$

where the indices ijk refer to the Cartesian components of the fields. The degeneracy factor D represents the number of distinct permutations of the frequencies ω_n and ω_m. In Eq. (28), the first frequency argument of the nonlinear susceptibility defines the type of nonlinear process. In the case of second-harmonic generation its value would be 2ω for instance. The second and third arguments of the susceptibility refer to the frequency of the two fields that are involved in the nonlinear process. As can be seen from Eq. (28), each component of the second-order nonlinear polarization is given by a sum of 9 terms, so that the corresponding second-order susceptibility tensor has 27 elements. Fortunately, these tensors possess some symmetry properties that can be invoked to reduce the number of independent elements. In particular, the tensor elements are unchanged by permutation of the last two indices and permutation of the second and third frequency arguments:

$$\chi_{ijk}^{(2)}(\omega_n + \omega_m; \omega_n, \omega_m) = \chi_{ikj}^{(2)}(\omega_n + \omega_m; \omega_m, \omega_n) \tag{29}$$

This property is referred to as intrinsic permutation symmetry and applies to all materials and experimental conditions. If restrictions are made, the tensor can be further simplified. For a loss-less medium, that is when the imaginary part of the complex susceptibility is zero, the tensors possess full permutation symmetry. In this case, all the frequency arguments can be interchanged as long as the Cartesian indices are changed as well. The sign of the frequency argument must be inverted when the first argument is interchanged with any of the two others:

$$\chi_{ijk}^{(2)}(\omega_l; \omega_n, \omega_m) = \chi_{jki}^{(2)}(-\omega_n; \omega_m, -\omega_l) = \chi_{kij}^{(2)}(-\omega_m; -\omega_l, \omega_n) \tag{30}$$

When the optical frequencies involved in the nonlinear interaction are far from any resonance, the frequency dependence (or dispersion) of the optical response of the material can be ignored. In this case, the tensor elements are unchanged by the permutation of all Cartesian indices without changing the frequency arguments:

$$\chi_{ijk}^{(2)}(\omega_n + \omega_m; \omega_n, \omega_m) = \chi_{ijk}^{(2)} = \chi_{ikj}^{(2)} = \chi_{jki}^{(2)} = \chi_{jik}^{(2)} = \chi_{kij}^{(2)} = \chi_{kji}^{(2)} \tag{31}$$

This approximation is referred to as Kleinman symmetry and is used very often in the description of electro-optic and photorefractive polymers.

In the case of second-harmonic generation, the second-order nonlinear susceptibility tensor elements are symmetric in their last two Cartesian indices and are
unchanged by the permutation of their second and third frequency arguments because they are identical. Thus, Eq. (28) can be rewritten in the simplified form

$$
\begin{pmatrix} P_X^{(2)} \\ P_Y^{(2)} \\ P_Z^{(2)} \end{pmatrix} = \begin{pmatrix} \chi_{XXX}^{(2)} & \chi_{XYY}^{(2)} & \chi_{XZZ}^{(2)} & \chi_{XYZ}^{(2)} & \chi_{XXZ}^{(2)} & \chi_{XXY}^{(2)} \\ \chi_{YXX}^{(2)} & \chi_{YYY}^{(2)} & \chi_{YZZ}^{(2)} & \chi_{YYZ}^{(2)} & \chi_{YXZ}^{(2)} & \chi_{YXY}^{(2)} \\ \chi_{ZXX}^{(2)} & \chi_{ZYY}^{(2)} & \chi_{ZZZ}^{(2)} & \chi_{ZYZ}^{(2)} & \chi_{ZXZ}^{(2)} & \chi_{ZXY}^{(2)} \end{pmatrix} \begin{pmatrix} E_X^2 \\ E_Y^2 \\ E_Z^2 \\ 2E_Y E_Z \\ 2E_X E_Z \\ 2E_X E_Y \end{pmatrix}
\tag{32}
$$

where for simplicity we have ignored the frequency arguments. Note that the
number of independent tensor elements is now reduced from 27 down to 18 elements. Due to the intrinsic permutation symmetry, the tensor elements $\chi_{ijk}^{(2)}$ can
be expressed in contracted form $\chi_{il}^{(2)}$ with only two indices but where the first index i takes the values 1, 2 or 3 corresponding to the three Cartesian coordinates,
and the second index l varies between 1 and 6. The values of l refer to the six different combinations of the indices j and k with the following convention:

l:	1	2	3	4	5	6
j, k:	1, 1	2, 2	3, 3	2, 3 or 3, 2	1, 3 or 3, 1	1, 2 or 2, 1

In the contracted notation the tensor gives

$$
\tilde{\chi}^{(2)} = \begin{pmatrix} \chi_{11}^{(2)} & \chi_{12}^{(2)} & \chi_{13}^{(2)} & \chi_{14}^{(2)} & \chi_{15}^{(2)} & \chi_{16}^{(2)} \\ \chi_{21}^{(2)} & \chi_{22}^{(2)} & \chi_{23}^{(2)} & \chi_{24}^{(2)} & \chi_{25}^{(2)} & \chi_{26}^{(2)} \\ \chi_{31}^{(2)} & \chi_{32}^{(2)} & \chi_{33}^{(2)} & \chi_{34}^{(2)} & \chi_{35}^{(2)} & \chi_{36}^{(2)} \end{pmatrix}
\tag{33}
$$

In addition to the symmetry properties of the nonlinear susceptibility elements,
group theory provides more means to reduce further the number of tensor elements. According to group theory, all materials can be classified in one of the possible 32 crystal classes. By invoking symmetry, some of the tensor elements vanish
for a given crystal class or point group symmetry. Poled polymers such as photorefractive polymers belong to the ∞mm symmetry group and their second-order
susceptibility tensor reduces to

$$
\tilde{\chi}^{(2)} = \begin{pmatrix} 0 & 0 & 0 & 0 & \chi_{15}^{(2)} & 0 \\ 0 & 0 & 0 & \chi_{15}^{(2)} & 0 & 0 \\ \chi_{31}^{(2)} & \chi_{31}^{(2)} & \chi_{33}^{(2)} & 0 & 0 & 0 \end{pmatrix}
\tag{34}
$$

with $\chi_{15}^{(2)} = \chi_{31}^{(2)}$ when Kleinman symmetry is valid, leading to only two independent tensor elements.

In some early studies of second harmonic generation, the second-order susceptibility tensor elements were described by nonlinear tensor elements d_{il} that are simply related to the second-order nonlinear susceptibility by

$$d_{il} = \frac{1}{2}\chi_{il}^{(2)} \tag{35}$$

3.2
Fundamentals of Electro-Optics

In the general expression of the second-order nonlinear polarization given by Eq. (28) in the previous section, the frequency arguments in the susceptibility are not limited to optical frequencies and can take the value zero. In this case, when one of the fields is a d.c. field (zero frequency or low frequency compared with optical frequencies), the nonlinear polarization in Eq. (28) describes the process in which the refractive index is modified by an applied electrical field. The nonlinear susceptibility tensor elements for this interaction are of the form $\chi_{il}^{(2)}(\omega;\omega,0)$. Such a process, referred to as the electro-optic effect, was known well before the beginning of nonlinear optics and was first discovered by Friedrich Pockels in 1893. In its broadest sense, the electro-optic effect was defined as the change of refractive index resulting from the application of a d.c. or low frequency electric field. When that index change is proportional to the applied field it is called the *linear electro-optic effect* or *Pockels effect*. When the index change is quadratic in the applied electric field it is referred to as the *quadratic electro-optic effect* or *Kerr effect*. The Kerr effect was discovered first in 1875 by John Kerr. Obviously, the field of electro-optics started well before Peter Franken's experiment of second-harmonic generation in 1961 that initiated the field of nonlinear optics. Unfortunately, the two fields use different starting points to describe the action of a d.c. field, so that a relationship needs to be established in order to connect the two fields. In the framework of nonlinear optics, the starting point for non-resonant processes is usually to develop the susceptibility tensor into a power series of the total electric field. In contrast, in the framework of electro-optics the refractive index of the material $n(E)$ is described as a function of the d.c. or low frequency electric field E_0. Since the index changes are small with E_0 the index is expanded in a Taylor's series about $E_0 = 0$:

$$n(E_0) = n(E_0 = 0) + \frac{dn}{dE_0}\Bigg)_{E_0=0} E_0 + \frac{1}{2}\frac{d^2n}{dE_0^2}\Bigg)_{E_0=0} E_0^2 + \dots \tag{36}$$

and is generally rewritten in the form

$$n(E_0) = n(E_0 = 0) - \frac{1}{2} n^3 r E_0 - \frac{1}{2} n^3 s E_0^2 + \dots \tag{37}$$

where the two coefficients r and s are defined by

$$r = -\frac{2}{n^3} \frac{dn}{dE_0}\bigg)_{E_0=0} \quad , s = -\frac{1}{n^3} \frac{d^2 n}{dE_0^2}\bigg)_{E_0=0} \tag{38}$$

The coefficients r and s are called the linear and quadratic electro-optic coefficient, respectively.

As discussed in Sect. 2.1.2, the index of an anisotropic medium is described by the index ellipsoid (Eq. 22). If the coordinate system is chosen such that the axes do not match with the principal symmetry axes of the crystal, the index ellipsoid is described by the more general expression [11]

$$\left(\frac{1}{n^2}\right)_1 X^2 + \left(\frac{1}{n^2}\right)_2 Y^2 + \left(\frac{1}{n^2}\right)_3 Z^2 + 2\left(\frac{1}{n^2}\right)_4 YZ +$$
$$2\left(\frac{1}{n^2}\right)_5 XZ + 2\left(\frac{1}{n^2}\right)_6 XY = 1 \tag{39}$$

The linear change in refractive index due to an arbitrary electric field $\mathbf{E_0}$ (E_{0X}, E_{0Y}, E_{0Z}) is defined by

$$\Delta\left(\frac{1}{n^2}\right)_i = \sum_{j=1}^{3} r_{ij} E_{0j} \tag{40}$$

or

$$\begin{pmatrix} \Delta\left(\frac{1}{n^2}\right)_1 \\ \Delta\left(\frac{1}{n^2}\right)_2 \\ \Delta\left(\frac{1}{n^2}\right)_3 \\ \Delta\left(\frac{1}{n^2}\right)_4 \\ \Delta\left(\frac{1}{n^2}\right)_5 \\ \Delta\left(\frac{1}{n^2}\right)_6 \end{pmatrix} = \begin{pmatrix} r_{11} & r_{12} & r_{13} \\ r_{21} & r_{22} & r_{23} \\ r_{31} & r_{32} & r_{33} \\ r_{41} & r_{42} & r_{43} \\ r_{51} & r_{52} & r_{53} \\ r_{61} & r_{62} & r_{63} \end{pmatrix} \begin{pmatrix} E_{0X} \\ E_{0Y} \\ E_{0Z} \end{pmatrix} \tag{41}$$

The index ellipsoid can also be expressed in a more convenient way as

$$\sum_{ij} \eta_{ij} x_i x_j = 1 \tag{42}$$

where we introduced the *impermeability* tensor $\tilde{\eta}$ defined as

$$E_i = \sum_j \eta_{ij} D_j \Rightarrow \tilde{\eta} = \tilde{\varepsilon}^{-1} \tag{43}$$

In this case, the electro-optic tensor elements r_{ijk} are defined by the following expression:

$$\eta_{ij} = \eta_{ij}(E_0 = 0) + \sum_k r_{ijk} E_{0k} + \dots \tag{44}$$

Since the dielectric tensor $\tilde{\varepsilon}$ is real and symmetric, its inverse $\tilde{\eta}$ is also real and symmetric, and the tensor $\tilde{r}$ must be symmetric in its first two indices. The electro-optic tensor elements as defined in Eqs. (40) and (41) are expressed in the contracted notation:

$$r_{ijk} = r_{hk} \quad \text{with} \quad \begin{pmatrix} h = 1 \text{ for } i, j = 1,1 \\ h = 2 \text{ for } i, j = 2,2 \\ h = 3 \text{ for } i, j = 3,3 \\ h = 4 \text{ for } i, j = 2,3 \text{ or } 3,2 \\ h = 5 \text{ for } i, j = 1,3 \text{ or } 3,1 \\ h = 6 \text{ for } i, j = 1,2 \text{ or } 2,1 \end{pmatrix} \tag{45}$$

Hence, Eq. (44) can be rewritten in the form

$$\tilde{\eta} = \tilde{\eta}_0 + \Delta\tilde{\eta} \quad \text{with} \quad \Delta\tilde{\eta} = \tilde{r} : E_0 \tag{46}$$

or also

$$\begin{pmatrix} \Delta\eta_1 \\ \Delta\eta_2 \\ \Delta\eta_3 \\ \Delta\eta_4 \\ \Delta\eta_5 \\ \Delta\eta_6 \end{pmatrix} = \begin{pmatrix} r_{11} & r_{12} & r_{13} \\ r_{21} & r_{22} & r_{23} \\ r_{31} & r_{32} & r_{33} \\ r_{41} & r_{42} & r_{43} \\ r_{51} & r_{52} & r_{53} \\ r_{61} & r_{62} & r_{63} \end{pmatrix} \begin{pmatrix} E_{0X} \\ E_{0Y} \\ E_{0Z} \end{pmatrix} \tag{47}$$

From the two descriptions of the Pockels effect in the frameworks of electro-optics and nonlinear optics, one can show that the electro-optic tensor elements and the second-order nonlinear susceptibility elements are related by

$$r_{ij} = -\frac{8\pi}{n^4} \chi_{ji}^{(2)} \tag{48}$$

In deriving Eq. (48) one ignores the anisotropy of the linear part of the dielectric constant and one assumes that Kleinman's symmetry is valid. Under these conditions, using group theory to reduce the number of non-vanishing tensor elements, the electro-optic tensor for poled polymers is given by

$$\tilde{r} = \begin{pmatrix} 0 & 0 & r_{13} \\ 0 & 0 & r_{13} \\ 0 & 0 & r_{33} \\ 0 & r_{13} & 0 \\ r_{13} & 0 & 0 \\ 0 & 0 & 0 \end{pmatrix} \tag{49}$$

Before we close this section, let us emphasize the importance of the degeneracy factor D in Eq. (28). This factor takes different values depending on the nature of the nonlinear optical process. For instance, in the case of second harmonic generation its value is unity, while for an electro-optic process it takes the value 2. Therefore, great care must be taken in reconciling experimental values of nonlinear coefficients measured using different experiments.

4
Molecular and Bulk Nonlinear Optics

4.1
Microscopic Theory: First and Second Hyperpolarizability

Like macroscopic systems, molecules exhibit second-order nonlinearities only if the centrosymmetry is broken. This can be achieved in conjugated molecules by connecting each end of the conjugated path with groups that have a different electronic affinity. In other terms, the symmetry can be broken by deforming the π electron distribution by attaching a donor-like group at one end and an acceptor-like group at the other end as shown schematically in Fig. 4. This class of molecules is referred to as *push-pull* molecules. Due to an excess of charge at the acceptor side, the molecule has a dipole moment in its ground state. The response of the

A ——————| π – conjugated bridge |———— D

Fig. 4. Schematics of push-pull molecules for electro-optic and photorefractive applications

molecule to an electric field, or its polarizability, depends strongly on the direction of the applied field with respect to the molecule: charge flow is favored towards the acceptor, while hindered towards the donor. This asymmetric polarization provides strong second-order nonlinear optical properties. On a molecular level, in the dipole approximation, the microscopic polarization components of a molecule under nonresonant excitation conditions can be written as

$$p_i = \mu_i + \sum_j \alpha_{ij} E_j + R^{(2)} \sum_{jk} \beta_{ijk} E_j E_k + R^{(3)} \sum_{jkl} \gamma_{ijkl} E_j E_k E_l + \dots \qquad (50)$$

where α_{ij}, β_{ijk}, γ_{ijl} are tensor elements and E_j, E_k, E_l are components of the electric field. The subscripts i, j, k, l refer to Cartesian coordinates expressed in the frame of the molecule. μ_i is the ground state dipole moment of the molecule. The linear polarizability is given by the tensor $\tilde{\alpha}$. The microscopic equivalent of the macroscopic susceptibility $\tilde{\chi}^{(2)}$ is described by $\tilde{\beta}$ and is called the *first hyperpolarizability*. Similarly, $\tilde{\gamma}^{(2)}$ is the second hyperpolarizability and is the molecular equivalent of the macroscopic susceptibility $\chi^{(3)}$. Note that $\tilde{\beta}$ is a *third-order* tensor that is at the origin of *second-order* nonlinear effects, and that is called *first* hyperpolarizability. In Eq. (50) $R^{(n)}$ are degeneracy factors that depend on the nonlinear process and the frequencies of the electric fields involved in the interaction.

4.2
Two-Level Model

With the structural flexibility of organic compounds, molecules with increasing nonlinearity could be synthesized and incorporated into polymer matrices. Today, numerous electro-optic polymers exhibit electro-optic coefficients that are higher than the inorganic crystal of lithium niobate ($LiNbO_3$) which is a standard in the electro-optics industry with an electro-optic coefficient of 30 pm/V. These advances were possible through the simultaneous improvement of the magnitude of the nonlinearity at a molecular level and through a better understanding of intermolecular interactions between molecules in the polymer matrix. To predict the nonlinear optical properties of push-pull molecules Oudar, Chemla, and others [12–15] proposed a simple two-level model in which they considered the intramolecular charge-transfer interaction between acceptor and donor, and the fact that β was not only governed by the ground-state electronic distribution but also by excited states. The hyperpolarizability was then taken to be the sum of two contributions:

$$\beta = \beta_{add} + \beta_{CT} \qquad (51)$$

where β_{add} is the additive part of the substituents and their individual interaction with the conjugated π-electron system, and β_{CT} is the contribution from the intramolecular charge-transfer interaction between acceptor and donor through the

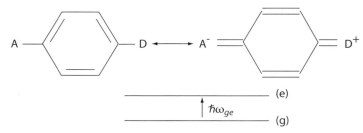

Fig. 5. Ground state and first excited state of a push-pull molecule

conjugated connecting bridge. The charge-transfer part is approximated to a two-level interaction between the ground state (g) and the first excited state (e) as illustrated in Fig. 5. Within this two-level model the charge-transfer contribution to the hyperpolarizability is given by

$$\beta_{CT}(\omega_3,\omega_2,\omega_1) = R\frac{e^2 f \Delta \mu_{ge}}{2m\hbar\omega_{ge}} \frac{\omega_{ge}^2(3\omega_{ge}^2 + \omega_1\omega_2 - \omega_3^2)}{(\omega_{ge}^2 - \omega_1^2)(\omega_{ge}^2 - \omega_2^2)(\omega_{ge}^2 - \omega_3^2)}$$

(52)

where R is a degeneracy factor, e is the elementary charge, m the electron mass, ω_i are the frequencies of the optical waves involved in the second-order nonlinear process, $\hbar\omega_{ge}$ is the energy difference between the ground state (g) and the excited state (e), $\Delta\mu_{ge} = \mu_e - \mu_g$ is the difference in dipole moments between the first excited state and the ground state, and f is the oscillator strength of the transition that is related to the *transition dipole moment* μ_{ge} through

$$f = \frac{2m}{\hbar^2}\hbar\omega_{ge}\mu_{ge}^2$$

(53)

Substituting for $\omega_3 = 2\omega$, $\omega_1 = \omega_2 = \omega$ in Eq. (52) leads to the following expression for the charge-transfer contribution to the hyperpolarizability for a second-harmonic generation process:

$$\beta_{CT}^{SHG}(2\omega,\omega,\omega) = \beta^{SHG}(0)\frac{\omega_{ge}^4}{(\omega_{ge}^2 - \omega^2)(\omega_{ge}^2 - (2\omega)^2)}$$

(54)

with

$$\beta^{SHG}(0) = \frac{3e^2}{\hbar^2}\frac{\Delta\mu_{ge}\mu_{ge}^2}{\omega_{ge}^2} \propto \frac{(\mu_e - \mu_g)\mu_{ge}^2}{E_{ge}^2}$$

(55)

where the superscript *SHG* refers to second harmonic generation. $\beta^{SHG}(0)$ is called *the dispersion free hyperpolarizability* since it does not contain any dependence on the frequency of the optical fields involved in the nonlinear interaction.

Similarly, starting from Eq. (52), the charge-transfer contribution to the hyperpolarizability of an electro-optic effect is obtained:

$$\beta_{CT}^{EO}(\omega,0,\omega) = \beta^{EO}(0)\frac{\omega_{ge}^2(3\omega_{ge}^2 - \omega^2)}{3(\omega_{ge}^2 - \omega^2)^2} \tag{56}$$

where the superscript *EO* refers to the electro-optic effect. When comparing values of the hyperpolarizability of different chromophores, the type of interaction (SHG or EO) should always be considered, as well as the wavelength at which it was measured. Note that due to the dispersion factors appearing in Eqs. (56) and (54), the nonlinearity is enhanced when the wavelength of the experiment is close to the wavelength of the charge-transfer absorption band. This is illustrated in Fig. 6 where the electro-optic hyperpolarizability normalized by the dispersion-free value is plotted vs wavelength for hypothetical chromophores with the same nonlinearity but with different linear absorption properties. The charge-transfer absorption band of the chromophore with energy $E_{ge} = \hbar\omega_{ge}$ is usually characterized by its corresponding wavelength $\lambda_{max} = hc/E_{ge}$.

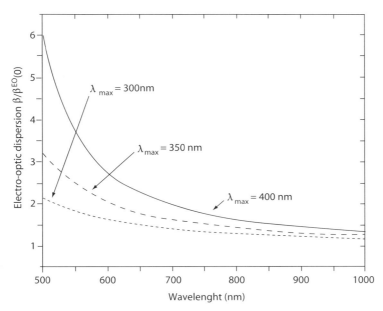

Fig. 6. Dispersion of the electro-optic hyperpolarizability for chromophores with different linear absorption properties

4.3
Bond-Length Alternation Model

While the two-level model guided the design of second-order nonlinear optical molecules for almost two decades, a model in which β is correlated to the degree of ground-state polarization was developed in recent years and has provided a way to synthesize molecules with larger nonlinearities [16]. The degree of ground-state polarization, or in other words the degree of charge separation in the ground state, depends primarily on the chemical structure (as for example, the structure of the π-conjugated system, or the strength of the donor and acceptor substituents), but also on its surroundings (as for example, the polarity of the medium). In donor-acceptor polyenes, this variable is related to a geometrical parameter, the bond-length alternation (BLA) which is defined as the average of the difference in the length between adjacent carbon-carbon bonds. To understand better the correlation between β and the degree of ground state polarization of the molecule, it is illustrative to discuss the wave function of the ground state $|\psi>$ in terms of a linear combination of the two limiting resonance structures:

$$|\psi>= a|n> +b|z> \tag{57}$$

where $|n>$ denotes the neutral form characterized by a positive BLA, and $|z>$ the charge-separated form characterized by a negative BLA (since the double and single bond pattern is now reversed relative to the neutral form) (see Fig. 7). The charge-separated form is also called *zwitterionic* or *quinoidal*. For substituted polyenes with weak donors and acceptors, the neutral resonance form dominates the ground-state wavefunction, and the molecule has a high degree of bond-length alternation. In other words, the coefficient a in Eq. (57) is much larger than b. With stronger donors and acceptors, the contribution of the charge-separated resonance form to the ground state increases and simultaneously, BLA decreases. When the two resonance forms contribute equally to the ground-state structure, the molecule exhibits essentially no BLA, corresponding to a situation where $a = b$ in Eq. (57). This zero BLA situation is called the *cyanine* limit, referring to the common structure of a cyanine molecule. Such cyanine molecules are known to be represented by two degenerate resonance forms, resulting in structures with virtually no BLA, as represented by the medium structure in Fig. 7. Finally, if the charge-separated form dominates the ground-state wave function, the molecule acquires a reversed bond-alternation pattern ($a \ll b$). BLA is thus a measurable parameter that is related to the *mixing* of the two resonance forms in the actual ground-state structure of the molecule.

Correlation between the molecular β with the ground-state polarization and consequently with BLA was demonstrated through quantum chemical calculations in polyene molecules [17, 18]. In these studies the ground-state polarization

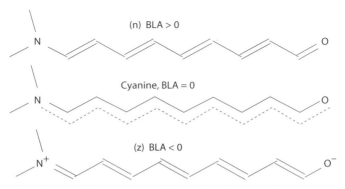

Fig. 7. Neutral, cyanine and charge-separated form of a polyene molecule and corresponding BLA value

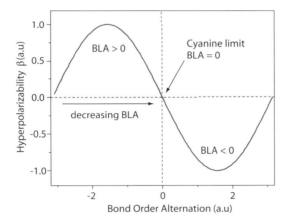

Fig. 8. Magnitude of β as a function of BLA or BOA

was tuned by applying an external static electric field of varying strength and β was correlated with the average of the difference in the π-bond order between adjacent carbon-carbon bonds (more simply denoted the Bond-Order Alternation, BOA) that is qualitatively related, as BLA, to the *mixing* of the two resonance forms in the actual ground-state structure of the molecule. The π-bond order is a measure of the double-bond character of a given carbon-carbon bond (a BOA of about -0.55 is calculated for a donor-acceptor polyene with alternating double and single bonds). Upon going from the neutral polyene limit to the cyanine limit, β first increases, peaks in a positive sense for an intermediate structure, decreases and passes through zero at the cyanine limit as shown in Fig. 8. From that limit and going to the charge-separated resonance structure, β continues to decrease and thus becomes negative, peaks in a negative sense, and then decreases again (in absolute value) to become smaller in the charge-separated structure. Evidence for the proposed relationships was derived from a study of a series of six molecules that spanned a wide range of ground-state polarization and therefore BLA, that were examined in solvents of varying polarity by electric field induced second harmonic generation (EFISH) experiments.

By applying this approach new molecules for electro-optic and photorefractive applications could be developed. For a review of the design of nonlinear molecules using the BLA model see for instance [19].

4.4
Oriented Gas Model

4.4.1
Electro-Optic Properties

So far, we have described the nonlinear optical properties of molecules on a microscopic level. Macroscopic nonlinearities in polymer films are obtained by doping or functionalizing the polymer with these molecules or molecular groups. To be electro-optic, a bulk material comprised of nonlinear molecules must also lack an inversion center if the molecular β is to lead to a macroscopic nonlinearity. The method most widely used to impart noncentrosymmetry in noncrystalline systems is the *poled-polymer* approach. If dipolar nonlinear species are dissolved in a polymer and subjected to a large electric field, at or above the glass transition temperature T_g of the polymer, the interaction of the dipole with the field causes the dipolar species to orient, to a certain extent, in the direction of the applied field. If the polymer is cooled back to the glassy state, with the field applied, then the field-induced noncentrosymmetric orientation can be frozen in place, yielding a material with a second-order optical nonlinearity. This process is referred to as *poling*.

To understand and optimize the electro-optic properties of polymers by the use of molecular engineering, it is of primary importance to be able to relate their macroscopic properties to the individual molecular properties. Such a task is the subject of intensive research. However, simple descriptions based on the oriented gas model exist [20, 21] and have proven to be in many cases a good approximation for the description of poled electro-optic polymers [22]. The oriented gas model provides a simple way to relate the macroscopic nonlinear optical properties such as the second-order susceptibility tensor elements expressed in the orthogonal laboratory frame {X,Y,Z}, and the microscopic hyperpolarizability tensor elements that are given in the orthogonal molecular frame {x,y,z} (see Fig. 9).

Because of its simplicity, the oriented gas model relies on a large number of simplifications and approximations: (i) at the poling temperature the chromophores are assumed free to rotate under the influence of the applied field and any coupling

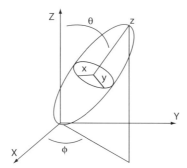

Fig. 9. Laboratory frame {X,Y,Z} and molecular frame {x,y,z}

or reaction from the surrounding matrix is ignored; (ii) the chromophores have cylindrical symmetry and the only nonvanishing hyperpolarizability tensor element is β_{zzz} where z is the symmetry axis of the chromophore; (iii) the permanent dipole moment μ of the chromophore is oriented along the z axis of the molecule; (iv) the chromophores are assumed independent and non-interacting. Under this last approximation, the bulk response of a material is given by the sum of the responses from its molecular components. The relationship between the macroscopic polarization components in the laboratory frame {X,Y,Z} and the molecular components can be written as

$$P_I(t) = \frac{1}{V}\left(\sum p_i(t)\right)_I \tag{58}$$

where V is the volume, the index I refers to the laboratory frame, and the index i to the frame attached to the molecule. Due to the large number of molecules, the summation in Eq. (58) is replaced by a thermodynamic average. The second-order nonlinear susceptibility tensor elements can be written as [23]

$$\chi_{IJK}^{(2)} = N\left\langle a_{Ii}a_{Jj}a_{Kk}\beta_{ijk}^{*}\right\rangle_{IJK} \tag{59}$$

where N is the density of chromophores, and the a_{Ij} are the director cosines or the projection of the new axis I (in the laboratory frame) on the former axis j (in the molecular frame). The brackets <> in Eq. (59) denote the orientational ensemble average of the hyperpolarizability tensor elements. The superscript * denotes that the local field correction factors have been included in β. By definition of a thermodynamic average, Eq. (59) can be rewritten as

$$\left\langle a_{Ii}a_{Jj}a_{Kk}\beta_{ijk}^{*}\right\rangle_{IJK} = \int a_{Ii}a_{Jj}a_{Kk}\beta_{ijk}^{*}G(\varphi,\theta,\psi)d\Omega \tag{60}$$

where $G(\varphi, \theta, \psi)$ is the normalized orientational distribution function. φ, θ, ψ are the Euler angles. $d\Omega = d\psi \sin\theta \, d\theta \, d\varphi$ is the differential solid angle. In the following, an additional simplification is made: the distribution function $G(\varphi, \theta, \psi)$ is assumed independent of φ and ψ. This assumption is related to the approximation that the chromophores have cylindrical symmetry and that the second-order non-linear optical properties depend mainly on the average polar angle θ of the molecules with the direction of the poling field (see Fig. 9).

The normalized orientational distribution function $G(\varphi, \theta, \psi)$ is approximated to a Maxwell-Boltzmann distribution function that gives

$$G(\Omega)d\Omega = \frac{\exp\left(-\dfrac{U(\theta)}{k_B T}\right)\sin\theta d\theta}{\displaystyle\int_0^\pi \exp\left(-\dfrac{U(\theta)}{k_B T}\right)\sin\theta d\theta} \tag{61}$$

where $U(\theta)$ is the interaction energy between the poling field and the dipole moment. This interaction energy is given by

$$U(\theta) = -\vec{\mu}^* \cdot E_p - \frac{1}{2} p \cdot E_p \approx -x k_B T \cos\theta \quad \text{with} \quad x = \frac{\mu^* E_p}{k_B T} \tag{62}$$

where μ^* is the modulus of the permanent dipole moment of the chromophore and is corrected for local field effects by the surrounding matrix. E_p is the modulus of the poling field that is applied along the Z axis of the laboratory frame. The second term in the right-hand side of Eq. (62) describes the reorientation of the molecule due to the induced dipole moment p. Since push-pull molecules have generally a high permanent dipole moment, it is a good approximation to neglect that second contribution. The force associated with the potential $U(\theta)$ is the torque exerted by the electric field on the molecule that tends to orient the molecule in the direction of the poling field. This torque tends to minimize the interaction energy, that is to reduce the value of the angle θ. With the notation introduced in Eq. (62) the orientational average of any function $g(\theta)$ that depends solely on the polar angle θ through the interaction potential $U(\theta)$, can be written as

$$\langle g(\theta)\rangle = \frac{\displaystyle\int_0^\pi g(\theta)\exp(x\cos\theta)\sin\theta d\theta}{\displaystyle\int_0^\pi \exp(x\cos\theta)\sin\theta d\theta} \tag{63}$$

According to Eq. (59), the two independent tensor elements for poled polymers when Kleinman symmetry is valid are given by

$$\chi^{(2)}_{ZZZ} = N\left\langle \left(\hat{K}\cdot\hat{k}\right)\left(\hat{K}\cdot\hat{k}\right)\left(\hat{K}\cdot\hat{k}\right)\right\rangle \beta^*_{zzz} = N\left\langle \cos^3\theta\right\rangle \beta^*_{zzz} \tag{64}$$

$$\chi_{ZXX}^{(2)} = N\left\langle \left(\hat{K}\cdot\hat{k}\right)\left(\hat{I}\cdot\hat{k}\right)\left(\hat{I}\cdot\hat{k}\right)\right\rangle \beta_{zzz}^* = N\left\langle \cos\theta\sin^2\theta\cos^2\varphi\right\rangle \beta_{zzz}^* \tag{65}$$

Replacing the ensemble average by Eq. (63), Eqs. (64) and (65) can be rewritten as

$$\chi_{ZZZ}^{(2)} = NL_3(x)\beta_{zzz}^* \tag{66}$$

$$\chi_{ZXX}^{(2)} = \frac{N}{2}\left(L_1(x) - L_3(x)\right)\beta_{zzz}^* \tag{67}$$

where we have introduced the Langevin functions $L_n(x)$ of order n and with an argument x defined as

$$L_n(x) = \frac{\int_0^\pi \cos^n\theta \exp(x\cos\theta)\sin\theta\,d\theta}{\int_0^\pi \exp(x\cos\theta)\sin\theta\,d\theta} \tag{68}$$

The Langevin functions for $n = 1$ to 3 are given by

$$L_1(x) = \coth x - \frac{1}{x}$$

$$L_2(x) = 1 + \frac{2}{x^2} - \frac{2}{x}\coth x \tag{69}$$

$$L_3(x) = \left(1 + \frac{6}{x^2}\right)\coth x - \frac{3}{x}\left(1 + \frac{2}{x^2}\right)$$

When $x = \mu^* E_p/kT$ is smaller than unity, these functions can be approximated to

$$L_1(x) \approx \frac{x}{3}, \quad L_2(x) \approx \frac{1}{3} + \frac{2x^2}{45}, \quad L_3(x) \approx \frac{x}{5} \tag{70}$$

The validity of this approximation for $L_3(x)$ for values of the argument $x < 1$ is illustrated in Fig. 10. The condition $x < 1$ is satisfied in most of the experimental conditions. For instance, for a molecule with a dipole moment of $\mu^* = 15$ D (or 15×10^{-18} esu) and a poling field of $E_p = 100$ V/μm (corresponding to 3333.33 statvolt/cm) at a temperature of 450 K, the value for the argument is $x = 0.8$. Thus, for small poling fields, when the linear approximation for the Langevin functions holds, the two independent susceptibility tensor elements given by Eqs. (66) and (67) give

$$\chi_{ZZZ}^{(2)} = N\frac{\mu^* E_p}{5k_B T}\beta_{zzz}^* \tag{71}$$

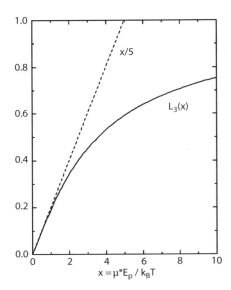

Fig. 10. Langevin function $L_3(x)$ and its linear approximation

$$\chi^{(2)}_{ZXX} = N \frac{\mu^* E_p}{15 k_B T} \beta^*_{zzz}$$

(72)

Note that within all the approximations made in that section we get the result that $\chi^{(2)}_{ZXX} = \chi^{(2)}_{ZZZ} / 3$.

4.4.2
Orientational Birefringence

Due to their rod-like shape, the chromophores possess a linear polarizability that is very different for directions of the optical field polarized parallel or perpendicular to the molecular axis. The poling process described in the previous section where the orientation of the molecules is changed by an external field leads, therefore, to birefringence in addition to the second-order nonlinear optical properties. In purely electro-optic polymers where poling is initially achieved by a spatially uniform dc field, the poling induces a permanent birefringence that is spatially uniform and plays generally only a minor role during electro-optic modulation. In contrast, in photorefractive polymers with a glass transition temperature close to room temperature, spatially modulated birefringence can be induced by the modulated internal space-charge field and can be of paramount importance since it can lead to strong refractive index gratings as will be shown later in Sect. 6.

The polarizability tensor of the molecule can be written in the molecular principal axes {x, y, z} as

$$
\tilde{\alpha} = \begin{pmatrix} \alpha_{xx} & 0 & 0 \\ 0 & \alpha_{yy} & 0 \\ 0 & 0 & \alpha_{zz} \end{pmatrix} = \begin{pmatrix} \alpha_\perp & 0 & 0 \\ 0 & \alpha_\perp & 0 \\ 0 & 0 & \alpha_{//} \end{pmatrix}
\tag{73}
$$

where the subscripts $\perp$ and $//$ refer to the direction perpendicular and parallel to the main z-axis of the molecule. In the linear regime and on a microscopic level, an optical field at frequency ω interacting with the polymer film will induce the molecular polarization

$$
p_i = \alpha_{ii} E_i(\omega)
\tag{74}
$$

where the index i refers to the different components in the frame of the molecule. The macroscopic polarization is given by the sum of the molecular polarizations. Since the molecules exhibit a strong anisotropic linear polarizability, the macroscopic polarization will depend on their average orientation. Since the average orientation can be changed by applying a poling field, the macroscopic polarizability can be continuously tuned by changing the value of the applied field. For a poling field applied along the Z-axis, the change in linear susceptibility along the poling axis between an unpoled and poled film is given by [24]:

$$
\Delta\chi_{ZZ}^{(1)} = N(\alpha_{//} - \alpha_\perp)^* \left[\left\langle \cos^2\theta \right\rangle - \frac{1}{3} \right]
\tag{75}
$$

where we have introduced the superscript $*$ to include field correction factors in the polarizability anisotropy. Recalling the relationship between the refractive index and the susceptibility $n^2 = 1 + 4\pi\chi$ (see Eq. 6), the index change associated with a change in linear susceptibility is given by

$$
\Delta n_{ZZ} = \frac{2\pi}{n} \Delta\chi_{ZZ}^{(1)}
\tag{76}
$$

and, therefore, for an optical field polarized along the poling axis Z the index change induced by the poling of the chromophores is given by

$$
\Delta n_Z = \frac{2\pi}{n} N(\alpha_{//} - \alpha_\perp)^* \frac{2}{45} \left(\frac{\mu^* E_p}{k_B T} \right)^2
\tag{77}
$$

where we have replaced the second-order Langevin function by its linear approximation (see Eq. 72).

The same derivation can be made for an optical field polarized along a direction perpendicular to the poling axis and the refractive index change is given by

$$\Delta n_X^{(1)}(\omega) = \Delta n_Y^{(1)}(\omega) = -\frac{1}{2}\Delta n_Z^{(1)}(\omega) \tag{78}$$

Note that according to Eqs. (77) and (78) the index change induced by the reorientation of anisotropic chromophores is quadratic in the poling field E_p and is therefore an orientational Kerr effect. The index change has also a quadratic dependence on the dipole moment. Equation (78) shows that the refractive index increases for an optical field polarized along the direction of the poling field and decreases for a polarization in a direction perpendicular to it.

5
Photorefractive Effect

The photorefractive effect refers to the field-induced change in refractive index of an optical material that results from a light-induced redistribution of electrons and holes. Photorefractivity can, therefore, be observed in materials that are photoconductive and that possess a field-dependent refractive index. The effect was discovered in the 1960s during second harmonic generation experiments in nonlinear crystals [25–28] and was rapidly recognized as a sensitive and reversible process to record and process holographic information. During the past decades it was studied [29–32] in inorganic crystals such as $LiNbO_3$, $BaTiO_3$, $KNbO_3$, $Bi_{12}SiO_{20}$, semiconductors such as GaAs, InP, CdTe [33], and semiconductor multiquantum wells (MQW) [34, 35]. In the 1990s, with the progress in molecular photonics, organic photorefractive materials were developed [36, 37]. Since the first demonstration of photorefractivity in a polymer [37], numerous polymer composites have been developed and the performance level of this new class of materials could rapidly be improved by several order of magnitude, to a level where they outperform numerous inorganic crystals that are expensive and difficult to grow. Current materials can exhibit millisecond response times and can have refractive index modulation amplitude close to 1%. Photorefractive polymers are truly multifunctional since they combine photoconductive, trapping, and electrooptic properties. In this section, we will review the fundamentals of photorefractivity and describe the basic physical model, the Kukhtarev model, that has been used to describe photorefractivity in inorganic materials [38, 39]. Polymers have optical and electrical properties that are quite different from those of periodic structures such as crystals and inorganic semiconductors in which transport for instance is described by band models. Nevertheless, so far the Kukhtarev model has provided a good framework to describe the photorefractive behavior of polymers. Models that account for the differences between amorphous organic materials and crystals remain the subject of active research [40].

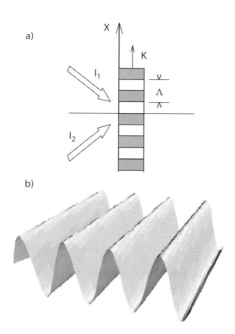

Fig. 11a,b. Illustration of the periodic light distribution generated by the interference of two coherent laser beams: **a** top view; **b** 3D view

Photorefractive materials are used to record thick phase gratings also called volume holograms. A phase grating is obtained through the periodic modulation of the refractive index of a material. It is called a phase grating as opposed to an amplitude grating because it results from the modulation of the real part of the refractive index and not its imaginary part which would correspond to the modulation of the absorption coefficient (see Eq. 13). Phase gratings are preferred to amplitude gratings because they can diffract an incident laser beam with a 100% efficiency. A hologram is recorded in an optical material by interfering two coherent laser beams as illustrated in Fig. 11. This interference pattern produces a spatially modulated intensity distribution $I(x)$ given by [41]

$$I(x) = I_0 \left[1 + m \cos(2\pi x / \Lambda) \right] \tag{79}$$

where $I_0 = I_1 + I_2$ is the total incident intensity, i.e., the sum of the intensities of the two beams and Λ is the grating spacing, i.e., the distance between two light maxima. In Eq. (79), m is the fringe visibility given by

$$m = \frac{2\sqrt{I_1 I_2}}{(I_1 + I_2)} \, \boldsymbol{e}_1 \cdot \boldsymbol{e}_2^* \tag{80}$$

where e_i are the unitary polarization vectors of the two interfering beams. The superscript * denotes the complex conjugate.

The different steps leading to the formation of a photorefractive grating are shown in Fig. 12. They comprise: (i) the absorption of light and the generation of charge carriers; (ii) the transport through diffusion and drift of electrons and holes over distances that are a fraction of the grating spacing leading to separation of holes and electrons; (iii) the trapping of these carriers and the build-up of a space-charge field; (iv) and the modulation of the refractive index by the periodic space-charge field.

With visible light, the period of the sinusoidal light distribution which is given by the value of the grating spacing can vary generally from a fraction of a micrometer to a couple of tens of micrometers. Following a spatially periodic photoexci-

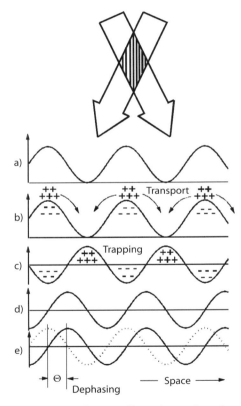

Fig. 12a–e. Illustration of the photorefractive effect. The overlap of two coherent laser beams creates an optical interference pattern (**a**). In the high intensity regions, charge carriers are generated (**b**). One type of carriers is transported and trapped (**c**), creating an alternating space-charge field (**d**). The space charge field induces a change in refractive index (**e**). The resulting index grating is phase shifted with respect to the initial light distribution

tation, more carriers are generated in the regions near the light maxima because the photogeneration rate can be assumed proportional to the local value of the optical intensity. In the simplest model, one majority carriers (electrons or holes) migrate from the positions of high intensity where they were generated and leave behind fixed charges of the opposite sign. In some materials, the minority carriers can migrate in the opposite direction but with a mobility that is much smaller than the mobility of the majority carriers. In traditional photorefractive crystals, the driving forces for carrier migration are diffusion due to concentration gradients and drift when a static electric field is applied. The migration process is limited by traps present in the material under the form of impurities or more generally any structural modification of the local environment of a site that leads to a deformation of the local potential and to a lower energy state for the carriers. The transport process being optically activated, the carriers can to some extent move away from the brighter regions to the darker regions where the conductivity is much lower and where they get trapped. The characteristic distance over which the carriers migrate in an efficient photorefractive material is, therefore, in the micrometer range. The carriers trapped in the darker regions and the fixed charged ions left behind in the brighter regions give rise to an inhomogeneous space-charge distribution. The space-charge distribution is in phase with the light distribution if transport is governed by diffusion only, and can be phase shifted when drift is present in response to an applied field. If the light is removed, this space-charge distribution can remain in place for a period of time which can vary between nanoseconds and years depending on the class of photorefractive materials. Next, this space charge distribution leads to a periodic electric field in the material that is 90° phase shifted with respect to the space-charge. The last step in the build-up of a photorefractive grating is the modulation of the refractive index by this internal space-charge field. In most of the photorefractive crystals this index modulation is produced via the electro-optic effect. However, any other field-dependent refractive index such as the orientational birefringence discussed in Sect. 4.4.2 can be used to produce photorefractive gratings. The spatial derivative that appears in Poisson's equation is of paramount importance for the photorefractive effect since it is the origin of the nonlocal response of photorefractive materials, namely the dephasing between the refractive index modulation and the initial light distribution. This phase shift is unique to photorefractive materials and can lead to energy transfer between two incident beams. It shows that the underlying meaning of the term photorefractivity is more than a refractive index variation induced by light and makes the photorefractive effect very different from many other mechanisms that can lead to grating formation such as thermal, chemical, and electronic nonlinearities. In contrast to many processes studied in nonlinear optics and described in other chapters of this book, the photorefractive effect is very sensitive and can be observed using milliwatts of laser power. Other unique properties of photorefractive materials are the possibility to store gratings during a long period of time but with the possibility to erase the grating at any time with a uniform light source. All these properties make photorefractive materials important for dynamic holographic recording/retrieval and all the related applications.

5.1
The Kukhtarev Model

An expression for the internal space-charge field can be obtained through the Kukhtarev model [38] that was developed to describe photorefractivity in most inorganic materials. In this model, the photorefractive material is described by a band model. As for a traditional semiconductor, the material consists of a conduction and a valence bands separated by a band gap as shown in Fig. 13. The model describes the transport of single carrier species and the band gap of the material contains localized energy levels that can be excited optically promoting either holes in the valence band (VB) or electrons in the conduction band (CB). In the model that we adopt here, we assume that the dopant is a donor with an energy level located in the band gap with concentration N_D^0. Furthermore, the crystal contains N_A acceptors with $N_A \ll N_D^0$ that are all ionized and that have accepted a charge from a donor. This leads to a concentration of ionized donors $N_D^+ = N_A^-$ prior to any optical excitation, leaving a concentration $N_D = N_D^0 - N_D^+$ of unionized donors. Optical absorption leads to the oxidation of the neutral donors with concentration N_D according to the following reaction:

$$N_D + h\nu \rightarrow N_D^+ + e \tag{81}$$

where $h\nu$ is the energy of the absorbed photon. The electron that is released in the conduction band (CB) undergoes transport until it gets trapped by an ionized donor to form a neutral donor according to the reaction

$$N_D^+ + e \rightarrow N_D \tag{82}$$

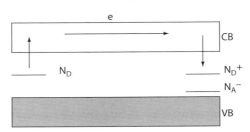

Fig. 13. Band model used to described photorefractivity using the Kukhtarev model

Under optical illumination, the concentration of ionized donors is given by the following rate equation:

$$\frac{\partial N_D^+}{\partial t} = (sI + \beta')(N_D^0 - N_D^+) - \gamma n_e N_D^+ \tag{83}$$

where s is a photo-ionization cross-section, β' a thermal generation rate, and γ a recombination coefficient. n_e is the density of electrons that is related to the current density j through the continuity equation

$$\frac{\partial n_e}{\partial t} = \frac{\partial N_D^+}{\partial t} + \frac{1}{e}\nabla \cdot j \tag{84}$$

where $\nabla \cdot$ is the divergence operator. The current equation is given by

$$j = n_e e \mu_e E^{(dc)} + \mu_e k_B T \vec{\nabla} n_e \tag{85}$$

where $\vec{\nabla}$ denotes the gradient. The field $E^{(dc)}$ is the static (or low frequency) electric field that develops within the material due to the combined effects of an applied field E_0 and any charge redistribution in the material. This field must satisfy Poisson's equation:

$$\varepsilon_{dc} \nabla \cdot E^{(dc)} = -4\pi e(n_e + N_A^- - N_D^+) \tag{86}$$

where ε_{dc} is the d.c. or low frequency dielectric constant. Equations (83)–(86) constitute a set of four differential equations with four unknowns: n_e, N_D^+, j, and $E^{(dc)}$. Since the optical excitation that is inducing the electric field is modulated, we can assume that all the unknown quantities will have the same periodic nature. Unfortunately, this system is nonlinear and does not have a general analytical solution. It can only be solved using numerical methods. However, zero- and first-order Fourier components of these distributions can be derived for steady-state conditions in the limit where the visibility of the fringe is small, that is for $m < 1$, where m is defined by Eq. (80). In this case, the unknown quantities can be approximated by the sum of a spatially uniform term and a term that has the spatial periodicity of the illumination:

$$E^{(dc)} = E_0 + \frac{1}{2}(E_1 e^{i2\pi x/\Lambda} + c.c.)$$

$$n_e = n_{e0} + \frac{1}{2}(n_{e1} e^{i2\pi x/\Lambda} + c.c.)$$

$$j = j_0 + \frac{1}{2}(j_1 e^{i2\pi x/\Lambda} + c.c.) \tag{87}$$

$$N_D^+ = N_{D0}^+ + \frac{1}{2}(N_{D1}^+ e^{i2\pi x/\Lambda} + c.c.)$$

For the electron-only transport conditions considered here, under weak illumination conditions and negligible dark conductivity, the Fourier coefficient of the first-order space-charge field at steady-state can be approximated to

$$E_1 = E_{sc} = -m \frac{E_q(iE_D + E_0)}{(E_q + E_D) - iE_0} \tag{88}$$

where E_D is the diffusion field and E_q is the limiting space-charge field. Note that we have changed the notation and replaced the first Fourier component E_1 by E_{sc} where the subscript sc refers to space-charge. The diffusion field is given by

$$E_D = \frac{2\pi k_B T}{\Lambda e} \tag{89}$$

and the limiting space-charge field by

$$E_q = \frac{2\Lambda e N_{eff}}{\varepsilon_{dc}} \tag{90}$$

where N_{eff} is the effective photorefractive trap density.

The space-charge field given by Eq. (88) is a complex number that is characterized by an amplitude and a phase. It can be rewritten as

$$E_{sc} = A + iB = |E_{sc}| e^{i\Theta} = \sqrt{A^2 + B^2}\, e^{i arctg B/A} \tag{91}$$

where A and B are the real and imaginary parts of the space-charge field, respectively. The amplitude of the space-charge field is given by

$$|E_{sc}| = m \left(\frac{(E_0^2 + E_D^2)}{(1 + E_D/E_q)^2 + (E_0/E_q)^2} \right)^{1/2} = m|E_m| \tag{92}$$

and the phase by

$$\Theta = arctg \left[\frac{E_D}{E_0} \left(1 + \frac{E_D}{E_q} + \frac{E_0^2}{E_D E_q} \right) \right] \quad \text{for } E_0 \neq 0 \tag{93}$$

$$\Theta = \frac{\pi}{2} \quad \text{for } E_0 = 0$$

Since the space-charge field produces the refractive index modulation, the phase Θ of the complex field represents the phase shift between the refractive index modulation and the initial periodic light distribution that generated the space-charge field.

Equation (88) describes the complex space-charge field obtained for a material in which electrons are the mobile charges. In hole transport materials, the same

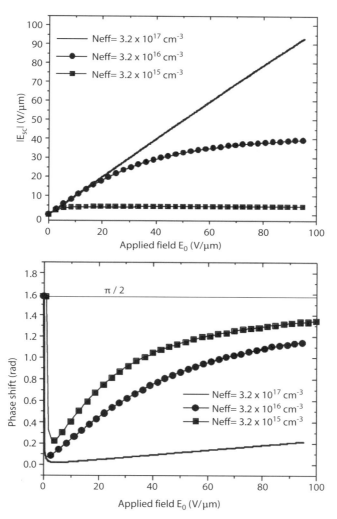

Fig. 14.a Amplitude of the space-charge field as a function of the value of the applied field calculated from Eq. (92) for different values of the photorefractive trap density N_{eff}. **b** Phase of the space-charge field as a function of the value of the applied field calculated from Eq. (93) for different values of the photorefractive trap density N_{eff}. For all calculations, the value of the grating spacing was $\Lambda = 3$ μm and the static dielectric constant was $\varepsilon_{dc} = 6.4$

derivation can be done and the resulting complex space-charge field is given by $E_{sc} = A - iB$ which leads to the same amplitude as for electron transport materials and a phase with opposite sign.

Examples of calculated values of the amplitude and the phase of the space-charge field as a function of applied field are shown in Fig. 14 for different values of the photorefractive trap density N_{eff}. Figure 14a shows that in the trap limited regime, when the density of traps is small, the space-charge field reaches a constant value. In contrast, when the trap density is large the space-charge field is equal to the applied field (here we assume that the applied field is applied along the grating vector for optimized drift conditions) and is saturated. Figure 14b illustrates how the phase shift is influenced by the trap density.

As discussed in Sect. 3.2, the internal periodic electric field described above will modulate the refractive index of an electro-optic material (see Eq. 37). The resulting index modulation can be written as

$$\Delta n = \frac{1}{2}\left(\Delta\tilde{n}e^{i2\pi x/\Lambda} + c.c.\right) \tag{94}$$

where $\Delta\tilde{n}$ is a complex amplitude that contains the phase information of the space-charge field given by

$$\Delta\tilde{n} = -\frac{1}{2}n^3 r_{eff}|E_{sc}|e^{i\Theta} \tag{95}$$

where r_{eff} is an effective electro-optic coefficient which depends on the orientation of the material and the polarization of the optical beams.

5.2
Four-Wave Mixing and Two-Beam Coupling

The periodic refractive index modulation given by Eq. (94) constitutes the thick phase hologram that leads to most of the applications with photorefractive materials. These applications, as will be discussed in more details in Sect. 7, can be divided into two categories depending on whether they are based on the diffraction of a beam on a grating that was previously recording with two independent light beams, or on the ability of two interfering beams to exchange energy during their propagation in the photorefractive medium. In the former case, the experimental geometry is similar to a four-wave mixing experiment (see Fig. 15a) and involves two writing beams and a reading beam that is counter-propagating with one of the

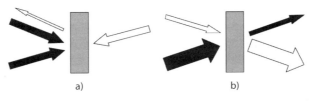

a)   b)

Fig. 15. a Geometry for four-wave mixing experiments. **b** Geometry for two-beam coupling experiments

writing beams. In this geometry, the reading beam is automatically Bragg matched with the grating induced by the two writing beams. The reading is generally not coherent with the pump beams and this type of experiment is often referred to as a light-induced grating experiment. In the case described in Fig. 15b, the two interfering beams interact during their propagation in the sample through the grating that they induce at the same time. This process is referred to as two-beam coupling.

In four-wave mixing experiments, the efficiency of the diffraction process is mainly governed by the amplitude of the index modulation. In contrast, in two-beam coupling experiments the amount of energy exchanged between the two beams is strongly affected by the value of the phase shift between the index modulation and the interference pattern produced by the two interacting beams. In other words, both amplitude and phase of the index modulation play an important role in two-beam coupling processes. Moreover, energy exchange in samples that are thick vanishes for a zero phase shift. Therefore, the observation of two-beam coupling in thick samples is often considered as a signature of the photorefractive nature of a grating since many other physical mechanisms that also lead to index modulations are local and do not exhibit any phase shift. One should mention, however, that energy exchange can be observed in thin samples due to coherent wave-mixing processes between several diffraction orders and should not always be assigned to an ultimate proof of photorefractivity in this case. To assess whether a grating is thin or thick, one should evaluate the value for the Q factor given by

$$Q = \frac{\pi \lambda d}{n \Lambda^2} \tag{96}$$

where d is the thickness of the grating. According to Kogelnik's coupled-wave theory [42] a grating can be considered thick for values of $Q > 5$.

For applications in which a hologram is recorded and retrieved in a four-wave mixing geometry, an important quantity is the diffraction efficiency η defined as the ratio between the intensity of the diffracted beam and the intensity of the incident reading beam. For thick sinusoidal phase gratings, this diffraction efficiency can be unity. According to Kogelnik's theory [42], the diffraction efficiency for a lossy thick phase grating can be approximated by

$$\eta = \exp\left(-\frac{\alpha d}{2}\left(\frac{1}{c_i} + \frac{1}{c_d}\right)\right)\frac{\sin^2(v^2 - \xi^2)^{1/2}}{(1 - \xi^2 / v^2)} \tag{97}$$

with

$$v = \frac{\pi \Delta n d}{\lambda(c_i c_d)^{1/2}}\hat{e}_i\hat{e}_d, \quad \xi = \frac{\alpha d}{4}\left(\frac{1}{c_i} - \frac{1}{c_d}\right) \tag{98}$$

where $\hat{e}_i$ and $\hat{e}_d$ are the polarization vectors of the incident reading beam and diffracted beam, respectively. In Eq. (97), Δn represents the modulus of the complex refractive index modulation amplitude described by Eq. (95). c_i and c_d are obliquity factors of the incident and diffracted beam, respectively. They are given by $c_i = \cos\alpha_1$ and $c_d = \cos\alpha_2$ where α_1 and α_2 are the angles of propagation of the incident and diffracted beams, respectively, measured inside the material with respect to the sample normal. Note that according to Eq. (97), the functional form of the diffraction efficiency is given by $\sin^2 x$ where the argument x is a linear function of the index modulation amplitude and the thickness of the grating. In samples that are sufficiently thick or in which the index modulation is such that the argument x takes values higher than $\pi/2$, the diffraction efficiency reaches a peak and decreases for larger values of d or Δn. In photorefractive polymers with a high dynamic range (high Δn) such effect is observed for numerous materials and is referred to as *overmodulation* of the diffraction efficiency.

In two-beam coupling experiments, the coherent energy exchange is characterized by a gain coefficient Γ that has units of cm^{-1}. Equations for the intensities of the two interacting beams propagating in the photorefractive material can be derived by solving the wave equation given by Eq. (5) for the total field given by the sum of the two beams with the slowly varying amplitude approximation, and by introducing the modulated refractive index given by Eq. (94). Simple algebra leads to the following coupled-wave equations for the intensities in the sample:

$$\frac{\partial I_1}{\partial z} = \Gamma \frac{I_1 I_2}{I_1 + I_2} - \alpha I_1 \tag{99}$$

$$\frac{\partial I_2}{\partial z} = -\Gamma \frac{I_1 I_2}{I_1 + I_2} - \alpha I_2 \tag{100}$$

with

$$\Gamma = \frac{4\pi}{\lambda \cos\theta} \Delta n_m \hat{e}_1 \hat{e}_2^* \sin\Theta \tag{101}$$

and

$$\Delta n_m = \frac{1}{2} n^3 r_{eff} |E_m| \tag{102}$$

where θ is the angle between the propagation direction of the beams inside the material and the normal to the sample. For this derivation, the geometry is such that the two beams are symmetric with respect to the normal of the sample (see Fig. 15b). The system of Eqs. (99) and (100) can be solved by defining the new variables $J(z) = I_1(z) + I_2(z)$ and $K(z) = I_1(z)/I_2(z)$ and with the boundary conditions $I_1(z = 0) = I_1(0)$ and $I_2(z = 0) = I_2(0)$. The solutions are given by

$$I_1(z) = \frac{I_1(0)(1+b)\exp(-\alpha z)}{1+b\exp(-\Gamma z)} \tag{103}$$

$$I_2(z) = \frac{I_1(0)b(1+b)\exp(-\alpha z)}{b+\exp(\Gamma z)} \tag{104}$$

where we have introduced the ratio $b = I_2(0)/I_1(0)$.

To determine the value of the gain coefficient defined by Eq. (101), one generally conducts the experiment in which the intensity of the transmitted beam $I_1(z)$ is first measured with the second beam, and then measured again with the second beam blocked. The ratio of these two intensity values is defined as

$$\gamma_0 = \frac{I_1(z)(I_2 \neq 0)}{I_1(z)(I_2 = 0)} = \frac{(1+b)\exp\Gamma z}{b+\exp\Gamma z} \tag{105}$$

Finally, the gain coefficient is derived from the measured value of γ_0 and the ratio b between the intensities of the two interacting beams:

$$\Gamma = \frac{1}{d}\Big[\ln(\gamma_0 b) - \ln(1+b-\gamma_0)\Big] \tag{106}$$

6
Organic Photorefractive Materials

While photorefractivity during the 1970s and 1980s was studied in many different classes of inorganic materials, organic photorefractive (PR) materials emerged only in the 1990s [36, 37]. The quest for organic photorefractive materials was driven by their electronic optical nonlinearity that leads to materials that combine high electro-optic coefficient and low dielectric constant, a property that improves the figure of merit for photorefractivity $Q' = n^3 r/\varepsilon_{dc}$, where n is the refractive index, r the electro-optic (or Pockels) coefficient, and ε_{dc} the low frequency dielectric constant. In the past decade, this new field of research experienced several milestones: a new and unexpected form of orientational photorefractivity was discovered [43]; consequently high dynamic range was obtained with low power semiconductor laser diodes; the polymers were successfully used in different applications, including holographic storage. With their plasticity and other unique properties, they constitute a strategic class of materials because they enable the mass production at low cost of new devices with low weight, and high performance. Thus, such materials are expected to highly impact dynamic holographic technologies.

Photorefractivity in an organic material was first reported in 1990 [36] by the ETH Zurich group in an organic single crystal. The same year a group at Eastman Kodak developed a multifunctional polymer that showed photoconductivity and

a)   b)

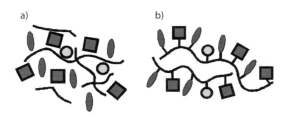

Fig. 16a,b. Schematics of different chemical designs for photorefractive polymers: **a** guest/host; **b** fully functionalized approach

electro-optic activity simultaneously [44]. Holographic recording and retrieval was not carried out in this material, but was demonstrated shortly after in another polymer composite at IBM Almaden [37]. While these two composites were obtained by doping an electro-optic polymer with charge transport molecules, another approach was developed independently at the University of Arizona [45] and at the University of Chicago [46], which consisted of the synthesis of a fully functionalized side-chain polymer with multifunctional groups. Several new polymers were synthesized at that time by several groups and it became rather obvious that the guest/host approach (see Fig. 16) was much faster and easier to implement and would provide a better way to test different combinations of polymers and molecules with photosensitivity, transport, and electro-optic activity.

A first milestone in this field occurred with the report by the IBM group of net gain and a diffraction efficiency of 1% in 125 μm-thick samples of the composite FDEANST:PVK:TNF [47]. Shortly after, the Arizona group reported on a composite based on the photoconductor PVK:TNF doped with the chromophore DMNPAA and that exhibited 6% diffraction efficiency in 100 μm-thick films [48]. That significant improvement was due to an additional plasticizer ECZ that was added to the composite and that was reducing the glass transition temperature enabling the orientation of the electro-optic chromophores with an applied field at room temperature. By improving the sample fabrication conditions, enabling higher fields to be applied, these composites exhibited nearly 100% diffraction efficiency, net gain coefficients of 200 cm^{-1}, and fully reversible index modulation amplitudes of 0.007 with a response time of 100–500 ms [49].

The report of these high diffraction efficiencies was very puzzling because the electro-optic properties of the dopant chromophores could not account for the high dynamic range that was observed. A possible explanation was proposed by the IBM group [43]. They attributed the high dynamic range to an orientational birefringence contribution caused by the spatially modulated orientation of the chromophores by the total field in the polymer. The observation of a change in coupling direction when changing the polarization of the beams from 'p' to 's' in DMNPAA:PVK:ECZ:TNF samples [49] was clear evidence that the dynamic range was indeed dominated by the orientational birefringence. The year 1994 was a turning point for the development of photorefractive polymers since it changed completely the design of chromophores: the Pockels effect was no longer the main driver but rather the orientational birefringence. A new form of photorefractivity was born: orientational photorefractivity where the index of a material is changed through the control of the orientation of optically anisotropic dopant molecules with a permanent dipole moment, by the internal space charge field that is produced like in a traditional photorefractive material by absorption followed by charge separation and trapping.

In this section, we will first provide a short discussion on some of the basics of the design of photorefractive polymers. Then we will review some of the basic physical properties such as photogeneration and transport in organic amorphous materials needed to develop and understand photorefractive polymers. We will describe the orientational photorefractive effect that is used in most of today's low glass transition temperature materials. Finally we will describe selected examples of photorefractive polymers and describe their optical and electrical properties.

6.1
Design of Photorefractive Polymers

As for inorganic materials, polymers must combine several key functionalities to exhibit photorefractive properties: weak absorption, transport of either positive or negative carrier species, trapping, and a field-induced refractive index modulation mechanism. Owing to the rich structural flexibility of organic molecules and polymers these functionalities can be incorporated into a given material in many different ways. The organic materials can be amorphous like polymers, or exhibit liquid crystalline phases, or can consist of nanocomposite phases such as polymer dispersed liquid crystals, or hybrid materials such as sol-gels. Photorefractivity has been evidenced in numerous classes of organic materials. With polymers, two main design approaches have been followed. (i) In the guest-host approach several materials with different functionalities are mixed into a polymer composite (see Fig. 16a). A polymer is sometimes used to provide the mechanical properties to the film. However, in some cases molecular materials form amorphous glasses by themselves and there is no need for a polymer compound. In these molecular glasses, the molecules can exhibit several functionalities simultaneously. (ii) In the fully functionalized approach, different functional groups are attached as side-groups to a polymer chain. The chain itself can in some cases incorporate one functionality (see Fig. 16b). This second approach requires a much larger synthetic effort and has not been used as much as the guest-host approach. Both approaches have their strengths and weaknesses. A major limitation of the guest-host approach is the compatibility of the different compounds that form the photorefractive composite. This compatibility limits the number of combinations of sensitizers, photoconductive and electro-active materials that can be tested. A significant problem in this approach is the phase separation between the different guest compounds that can occur over time. To be practical, a photorefractive material must combine good electrical and optical properties, but at the same time must exhibit high photostability under illumination, long shelf lifetime with high optical clarity, and good dielectric properties under extended bias of several kilovolts. Early guest-host photorefractive polymers had limited shelf lifetimes, but in recent years materials with shelf lifetimes of several years could be demonstrated [50–52]. Nevertheless, shelf lifetime should always remain a major concern when new materials are designed and characterized.

With the development of guest-host photorefractive polymers, the plasticizing effect of some of the guest molecules is used to tune the glass transition temperature (T_g) close to room temperature. In such materials, the orientational photorefractive effect is dominating and leads to high refractive index modulation amplitudes. There is, therefore, a clear distinction between materials that have a T_g that is well above the operation temperature (so-called high T_g polymers) and the ones that have a T_g that is only 10–20 °C above room temperature. In the former, the polymer is pre-poled at elevated temperatures and exhibit stable and spatially uniform electro-optic properties. In this case, the photorefractive effect is similar to the one observed in inorganic crystals and the refractive index modulation amplitude is given by Eq. (94). In low T_g materials, the situation is quite different and the orientational photorefractive effect has to be considered. This effect will be described in more details in Sect. 6.2.3. Most of the photorefractive polymers we will describe in Sect. 6.3 are low T_g materials.

As will become more obvious in the following sections, an applied electric field must be applied to photorefractive polymers. This is done by placing a thick

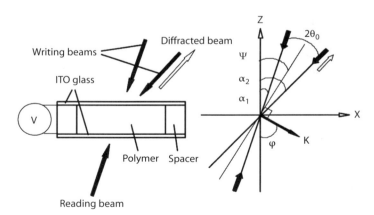

Fig. 17. Schematic of the tilted configuration for four-wave mixing and two-beam coupling experiments in photorefractive polymers

(≈ 100 µm) film of the polymer between two transparent conducting glass slides such as indium tin oxide (ITO) coated glass slides as shown in Fig. 17. The sample must have high optical quality and the film must exhibit high dielectric strength such that fields of several tens of V/µm can be applied to the sample without causing any dielectric breakdown. Therefore, starting materials with high purity must be used when preparing photorefractive polymers. The purity of many commercial products is not good enough to prepare good samples and additional purification steps should be carried out.

To give holograms efficiently in such samples, the geometry is such that the bisector of the two writing beams is tilted with respect to the normal of the sample. This way, the field applied to the two ITO coated glass slides has a non-vanishing component along the grating vector K of the interference pattern. In this tilted configuration, the field applied to the sample acts as a drift source for the carriers (see Fig. 17).

6.2
Physical and Chemical Properties of Photorefractive Polymers

In the sub-sections below, we briefly summarize the main physical and chemical properties of photorefractive polymers. Their properties differ from those of inorganic photorefractive materials in many different ways: (i) the basic properties such as photogeneration, transport and index modulation are different; (ii) polymers are amorphous materials as opposed to crystalline for most of the inorganic photorefractive materials; (iii) the mechanism for index change is different as discussed in the previous section. Each of the first two sub-sections below, photogeneration and transport in organic materials, are the subject of intense research across several fields and application areas. These complex phenomena are not well understood yet in monolithic systems and their understanding in multifunctional materials such as photorefractive polymers is an even bigger challenge because the different functional groups interact with each others. Such challenges are the subject of current and future research in photorefractive polymers. For completeness, we just provide below a brief conceptual overview and for more information we refer the reader to recent reviews in the field [53, 54].

6.2.1
Photogeneration

The space-charge build-up process can be divided into two steps: the electron-hole generation process followed by the transport of one carrier species. The creation of a correlated electron-hole pair (or exciton) after absorption of a photon can be followed by recombination. This process limits the formation of free carriers that can participate in the transport process and is, therefore, a loss for the formation of the space-charge. These properties result in low photogeneration efficiency unless an electric field is applied to counteract recombination. The quantum efficiency for carrier generation is, therefore, strongly field-dependent and increases with the applied field. A theory developed by Onsager [55, 56] for the dissociation of ion pairs in weak electrolytes under an applied field has been found to describe reasonably well the temperature and field dependence of the photogeneration efficiency in some of the organic photoconductors [54]. However, this model suffers from numerous deficiencies and provides only a qualitative functional dependence of the applied field. For instance, the thermalization radius used to obtain the best theoretical fit to the experimental data are clearly unphysical.

6.2.2
Transport

After photogeneration of free carriers, the next step in the build-up of a space-charge is their transport from brighter regions of the interference pattern, where they are generated, to the darker regions, where they get trapped. In contrast to inorganic photorefractive crystals with a periodic structure, photorefractive polymers have a nearly amorphous structure. The local energy level of each molecule/moiety is affected by its nonuniform environment. The disorder in amorphous photoconductors splits the conduction bands of molecular crystals into a distribution of localized electronic states. As a result, transport can no longer be described by band models but is attributed to intermolecular hopping of carriers between neighboring molecules or moieties [57–59].

In recent years, a model based on disorder due to Bässler and co-workers describes the transport phenomena of a wide range of different materials and emerged as a solid formalism to describe the transport in amorphous organic materials [60–67]. This so-called disorder formalism is based on the assumption that charge transport occurs by hopping through a distribution of localized states with energetical and positional disorder. So far, most of the predictions of this theory agree reasonably well with the experiments performed in a wide range of doped polymers, main-chain and side-chain polymers and in molecular glasses. However, when these photoconductors are doped with highly polar dyes such as photorefractive chromophores, significant differences have been observed that could not be rationalized within existing theories. More research is required to understand transport in photorefractive polymers.

In the Bässler formalism, disorder is separated into diagonal and off-diagonal components. Diagonal disorder is characterized by the standard deviation σ of the Gaussian energy distribution of the hopping site manifold (energetical disorder) and the off-diagonal component is described by the parameter Σ that describes the amount of positional disorder. Results of Monte-Carlo simulations led to the following universal law for the mobility when $\Sigma \geq 1.5$:

$$\mu(E,T)=\mu_0\exp\left[-\left(\frac{2}{3}\frac{\sigma}{k_BT}\right)^2\right]\exp\left\{C\left[\left(\frac{\sigma}{k_BT}\right)^2-\Sigma^2\right]E^{1/2}\right\} \tag{107}$$

where μ_0 is a mobility prefactor and C an empirical constant with a value of 2.9×10^{-4} (cm/V)$^{1/2}$. Equation (107) is valid for high electric fields (a few tens of V/μm) and for temperatures $T_g > T > T_c$ where T_g is the glass transition temperature and T_c the dispersive to nondispersive transition temperature. For a more detailed description the reader is referred to the original work by Bässler [60] and the recent review by Borsenberger [54].

Numerous simulations and experimental work showed that the width of the density of states σ in Eq. (107) is strongly dependent on the polarity of the polymer host and the dipole moment of the dopant molecule. It was found that the total width σ was comprised of a dipolar component σ_D and an independent Van der Waals component σ_{VdW} [68]:

$$\sigma^2=\sigma_{VdW}^2+\sigma_D^2 \tag{108}$$

Another important parameter that affects hole transport in guest/host systems is the relative position of the HOMO (highest occupied molecular orbital) and LUMO (lowest unoccupied molecular orbital) energy levels of the different dopant moieties. Charge transfer reactions involved in photogeneration and in transport, as well as trapping strongly depend on the relative positions of the frontier orbitals of the dopant molecules. Recent studies, for instance have shown that the density of traps can be controlled by engineering the HOMO level of the chromophore [69, 70].

6.2.3
Orientational Photorefractivity

As discussed in Sect. 5, in traditional electro-optic materials the refractive index modulation is induced by the space-charge field (see Eqs. 94 and 95) through the Pockels effect. Early photorefractive polymers were designed to mimic inorganic crystals and semiconductors. In these materials, a spatially uniform electro-optic activity was obtained by a poling process prior to the formation of photorefractive gratings. However, as photorefractive polymers were developed through the guest host approach the glass transition temperature T_g of the resulting composite was lowered: the electro-active chromophore was acting as a plasticizer. The addition of plasticizers reduced the T_g even further. If the operating temperature is close enough to T_g (around 10° below), the dipolar chromophores are free to reorient in the total electric field, that is the superposition of the applied field and the spatially modulated space-charge field. For each coordinate along the grating (X direction in Fig. 18a), the total field is changing both its amplitude and direction because the

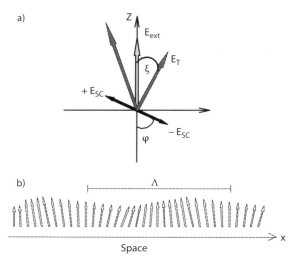

Fig. 18a,b. Schematic of: **a** the total poling field; **b** the spatially modulated orientation of the dipolar molecules in the sample

amplitude of the space-charge field is oscillating between its extrema $+E_{sc}$ and $-E_{sc}$. This results in a periodic orientation of the dipolar chromophores that orient in the total electric field as illustrated in Fig. 18b. As discussed in Sect. 4.4.2, the periodic orientation of anisotropic chromophores leads to a periodic refractive index modulation through orientational birefringence. In this case, the refractive index modulation has an electro-optic and a birefringent contribution. The former is directly related to the second-order nonlinear optical properties of the chromophore, while the latter is due to the anisotropy of the linear polarizability. To maximize the electro-optic effect, a chromophore should have a high hyperpolarizability β and simultaneously a high dipole moment μ. A high dipole moment μ is required to obtain a good orientation of the chromophores during the poling process through the torque exerted by the field on the chromophore (see Eqs. 62, 71 and 72). To maximize the orientational birefringence, a high polarizability anisotropy $\Delta\alpha = (\alpha_{//} - \alpha_{\perp})$ is desired. A high dipole moment μ is even more important in this case because the index change is quadratic in dipole moment (see Eq. 77). Based on the oriented gas model, one can define the following figure of merit for chromophores for orientational photorefractivity [71, 72]:

$$Q_{OP} = N_{max}(A(T)\Delta\alpha\mu^2 + \beta\mu) \tag{109}$$

where N_{max} is the maximum density of chromophores that can be doped or functionalized in the low glass transition polymer composite before chromophore aggregation or interaction occurs, and $A(T)$ is a scaling factor. The subscript OP in Q_{OP} stands for *orientational photorefractivity*.

The derivation of expressions for the electro-optic and birefringent contributions to the refractive index modulation in the case of orientational photorefractivity is more complex than for traditional photorefractivity. The tensors for the linear polarizability and the first hyperpolarizability are diagonal in the frame attached to the chromophores. In the case of a spatially uniform poling with the poling field applied along the laboratory Z axis, the molecular axis z usually coincides with the laboratory Z axis after poling. Here, the poling direction is changing for each coordinate along the X direction (i.e., along the grating) (see Fig. 18) and makes an angle $\xi(X)$ with the laboratory Z axis [43, 73]. By applying matrix transformations and by retaining only terms that have the spatial frequency K, the following expressions are derived for the different contributions to the refractive index modulation for each polarization of the reading beam. For the orientational birefringence contribution one gets

$$\Delta n_{K,s}^{(1)} = -\frac{2\pi}{n} B E_{ext} |E_{sc}| \cos\varphi \tag{110}$$

$$\Delta n_{K,p}^{(1)} = \tag{111}$$

$$\frac{2\pi}{n} B E_{ext} |E_{sc}| \left[2\cos\varphi \sin\alpha_1 \sin\alpha_2 - \cos\varphi \cos\alpha_1 \cos\alpha_2 + \frac{3}{2}\sin\varphi \sin(\alpha_1 + \alpha_2) \right]$$

with

$$B = \frac{2}{45} N(\alpha_{//} - \alpha_\perp)^* \left(\frac{\mu^*}{k_B T} \right)^2 \tag{112}$$

where the angles φ and α_i are defined in Fig. 17. For the electro-optic contribution the expressions are

$$\Delta n_{K,s}^{(2)} = \frac{8\pi}{n} C E_{ext} |E_{sc}| \cos\varphi \tag{113}$$

$$\Delta n_{K,p}^{(2)} = \tag{114}$$

$$\frac{8\pi}{n} C E_{ext} |E_{sc}| \left[\cos\varphi \cos\alpha_1 \cos\alpha_2 + 3\cos\varphi \sin\alpha_1 \sin\alpha_2 + \sin\varphi \sin(\alpha_1 + \alpha_2) \right]$$

with

$$C = \frac{N\beta^* \mu^*}{15 k_B T} \tag{115}$$

Note that the field E_{ext} in Eqs. (110), (111), (113), and (114) represents the total field applied between the ITO electrodes. Its component along the grating vector (which is represented by E_0 in Eqs. 88, 92, and 93) is given by $E_0 = E_{ext} \cos\varphi$.

All the orientational photorefractive properties discussed in this section are contingent on the ability of the chromophores to orient at the operation temperature, that is when the temperature is close to or above the glass transition temperature T_g. Most efficient photorefractive polymers to date are based on such orientational photorefractivity. However, polymers with high T_g can be designed in which poling (orientation of the chromophores) is achieved at high temperature and frozen in the material, leading to a spatially uniform electro-optic coefficient. In this case, photorefractivity is described by the Pockels effect as in inorganic photorefractive crystals.

6.3
Examples of Organic Photorefractive Materials

In this section we will describe a few examples of photorefractive polymer composites that illustrate the flexibility in the design of such materials. Organic photorefractive materials can be divided into several classes: (i) crystals, (ii) guest/host systems, (iii) fully functionalized polymers, (iv) glasses, (iv) sol-gels and organic/inorganic hybrid materials, (v) liquid crystals, (vi) polymer stabilized and polymer dispersed liquid crystals. Each of these classes contain sub-classes. For instance guest/host systems can be based on: (i) an inert polymer binder or (ii) a photoconducting polymer binder. In addition, any of the functional molecules and polymers can have more than one functionality. For instance, some chromophores can have electro-active and photoconducting properties simultaneously. Owing to the numerous organic photorefractive materials that have been reported in the 1990s, we will not present here an exhaustive list of all the materials with their performance. For more information the reader is referred to some recent reviews of the field [2, 3, 74] and the references cited therein. Instead we will present examples of the components used to fabricate low T_g photorefractive polymers: electro-active chromophores, sensitizers, plasticizers, and polymers.

Figure 19 shows examples of chromophores used in photorefractive polymers. DMNPAA and DEANST are widely used in photorefractive polymers and have been initially selected for their electro-optic properties. DMNPAA also exhibits high polarizability anisotropy and was used in the composites in which overmodulation of the diffraction efficiency could be observed in 105 μm-thick samples at applied fields of 60 V/μm, with net gain coefficients of 200 cm^{-1}. After the discovery of orientational photorefractivity, the design of chromophores was no longer driven by the electro-optic effect but rather by the orientational birefringence. DHADC-MPN and ATOP (chromophores (c) and (e) in Fig. 19) are examples of molecules that are close to the cyanine limit and have a high figure of merit (see

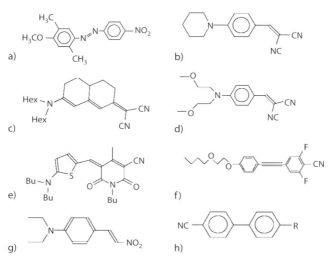

Fig. 19a–h. Examples of chromophores used in photorefractive polymers: **a** 2,5-dimethyl-4-(*p*-nitrophenylazo)anisole (DMNPAA); **b** 4-piperidinobenzylidene malononitrile (PD-CST); **c** 2-*N*,*N*-dihexyl-amino-7-dicyanomethylidenyl-3,4,5,6,10-pentahydronaphthalene (DHADC-MPN); **d** 4-di(2-methoxyethyl)aminobenzylidene malononitrile (AODCST); **e** amino-thienyl-dioxocyano-pyridine (ATOP) derivative; **f** fluorinated cyano-tolane chromophore (FTCN); **g** diethylamino-nitrostyrene (DEANST); **h** cyanobiphenyl derivative R = C_5H_{11} (5CB), R = OC_8H_{17} (8OCB)

Eq. 109) for orientational photorefractivity. To date, photorefractive polymers based on these materials exhibit the highest dynamic range at the lowest applied field [51, 75]. Figure 20 shows the diffraction efficiency and gain coefficient in a sample doped with DHADC-MPN that has a shelf lifetime of several years. Such samples stored for 6 days at 70 °C did not show any signs of phase separation, illustrating that guest/host photorefractive polymers with high stability can be fabricated. However, despite their high dynamic range, polymers based on these chromophores have a rather slow response time (seconds). Photorefractive polymers with millisecond response times have been designed by using PDCST and AODCST (chromophores (b) and (d) in Fig. 19) [76]. Such molecules have fast orientational dynamics in a plasticized PVK matrix. The photorefractive speed in such samples is limited by the photoconducting properties of the composite. These chromophores are colored and cannot be used across the full visible spectral region. Another chromophore that leads to millisecond response times when doped in a plasticized PVK matrix is FTCN (molecule (f) in Fig. 19) [77]. This molecule is transparent in most of the visible spectrum but has lower efficiency compared with PDCST and AODCST. To fabricate photorefractive polymers containing transparent electro-active molecules, 5CB (molecule (h) in Fig. 19), a molecule

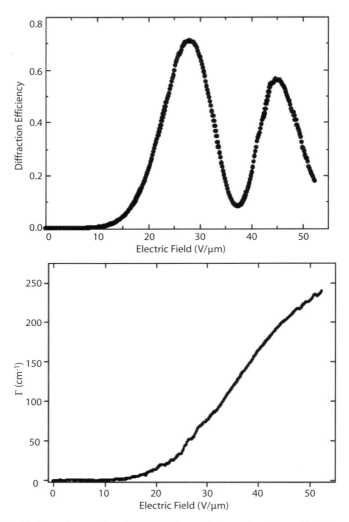

Fig. 20a,b. Field dependence of: **a** the diffraction efficiency; **b** gain coefficient measured at 633 nm in a 105 μm-thick sample of DHADC-MPN:PTCB:DIP:TNFDM (37.6:49.7:12.5:0.18 wt%)

that is known for its liquid crystalline phase, has been used as a dopant in PVK [78].

To tune the sensitivity of a photorefractive polymer to a particular spectral region, the composite is doped with a sensitizer. Examples of widely used sensitizers are shown in Fig. 21. TNF was used in numerous polymers based on the photoconducting polymer PVK because mixtures of PVK/TNF were used in the first

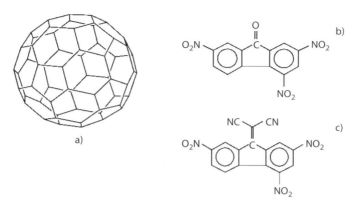

Fig. 21a–c. Examples of sensitizers: **a** fullerene C$_{60}$; **b** 2,4,7-trinitrofluorenone (TNF); **c** 2,4,7-trinitro-9-fluorenylidene-malononitrile (TNFDM)

commercial photocopiers and were extensively studied in the literature [54]. In most cases, the sensitizer itself does not absorb the light but rather the charge-transfer complex that is formed between the sensitizer and the molecule that enables charge transport. This charge-transfer complex exhibits a new absorption band that is different from any of the absorption of the two compounds. When the polymer composite is p-type, that is when the transport molecules have a donor-like character, the sensitizer is often a molecule with a strong acceptor character. In the case of an electron transport matrix, the sensitizer can be a donor-like molecule. Charge-transfer complexes are good sensitizers for photorefractive applications because they have a high photogeneration efficiency. Efficiencies of 100% have been reported recently in charge-transfer composites formed between triphenyldiamine derivatives and C$_{60}$ (molecule (a) in Fig. 21) [79]. The central wavelength of the charge-transfer band that be tuned by either changing the electron affinity of the acceptor-like molecule or the ionization potential of the donor-like molecule that form the complex. All molecules shown in Fig. 21 form a charge-transfer complex with PVK and often with some of the dopant electro-active molecules. While TNF and C$_{60}$ provide sensitivity in the red part of the spectrum, in particular at 633 nm that is widely used through the use of He:Ne lasers and semiconductor laser diodes, TNFDM enables near infra-red applications [80, 75].

Examples of polymers used in photorefractive applications are shown in Fig. 22. PVK and PSX both contain the carbazole moiety that has a donor-like character and provides for hole transport. Carbazole and some other hole transporting moieties have been widely used in the field of photorefractive polymers. The polarity of carbazole makes PVK a good matrix for a lot of the guest electro-active molecules. The choice of a polymer matrix is often dictated by its compatibility with

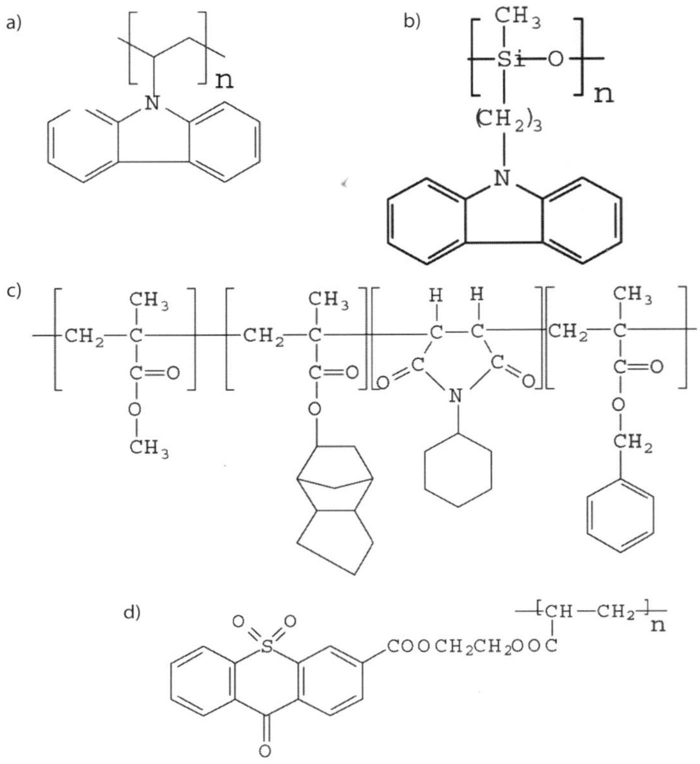

Fig. 22a–d. Examples of polymers used in photorefractive composites: **a** poly(vinylcarbazole) (PVK); **b** carbazole substituted polysiloxane (PSX); **c** poly(methylmethacrylate-*co*-tricy-clodecylmethacrylate-*co*-N-cyclohexylmaleimide-*co*-benzyl methacrylate) (PTCB); **d** thioxanthene containing polymer (P-THEA)

some of the other functional molecules that form the photorefractive composite. Hence, inert polymers have been used in some instances. PTCB (polymer (c) in Fig. 22) has an excellent compatibility with DHADC-MPN (see Fig. 19c) and good optical quality. The lack of transport properties however is responsible for the slow response time of such samples. Other inert binders that have been used include poly(methylmethacrylate) (PMMA) [81, 82], poly(vinylbutyral) (PVB), polystyrene (PS), the polyimide Ultem [82], bisphenyl-A-polycarbonate (PC) [83], and poly(*n*-butyl methacrylate) (PBMA) [84]. Recently, an electron transporting photorefractive polymer (P-THEA) was reported [85]. The electron-transporting nature of the composite was confirmed by a reverse two-beam coupling compared with PVK-based materials. Electron transport was achieved through the thioxanthene unit (see polymer (d) in Fig. 22).

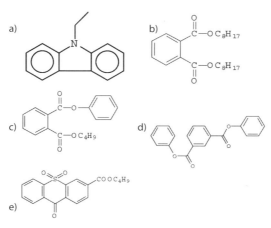

Fig. 23a–e. Examples of plasticizers: **a** *N*-ethylcarbazole (ECZ); **b** diisooctyl phthalate (DOP); **c** benzyl butyl phthalate (BBP); **d** diphenyliso phthalate (DIP); **e** 9-oxo-9*H*-thioxanthene-3-butyloxycarbonyl-10,10-dioxide (TH-*n*Bu)

In the photorefractive polymer composites described above, the value of the glass transition temperature T_g is of great importance because it needs to be in a range where the electro-active chromophores can be reoriented by the photorefractive space-charge charge field. Adjustment of T_g is done though the addition of a plasticizer. Examples of plasticizers are shown in Fig. 23. As for the polymer matrixes, plasticizers can be photoconducting or inert. ECZ (molecule (a) in Fig. 23) is widely used with PVK. Inert plasticizers such as BBP (molecule (c) in Fig. 23) have also been combined with PVK [86].

In an effort to optimize simultaneously the dynamic range and the response time in a given material, an attractive design concept is to have an amorphous glass of a multifunctional molecule that combines photosensitivity, transport, trapping, and electro-activity. This way the transport properties do not get diluted when a large amount of electro-active dopant is used, and vice-versa. In addition, the material should be processable into thick films that have a good shelf lifetime. Photorefractive glasses were first developed with the bifunctional chromophore 2BNCM (molecule (a) in Fig. 24) [87]. These chromophores combine high birefringence with transport properties but unlike previous chromophores they do not crystallize. The glasses that are formed have high optical quality and long stability. Photorefractivity was observed in these glasses and the refractive index modulation amplitude was 1.5 times higher than that observed in DMNPAA-based photorefractive polymers. However, the speed of these materials was quite slow (of the order of a minute). Another monolithic approach based on multifunctional carbazole trimers (molecule (e) in Fig. 24) was reported [88]. The trimers have a low

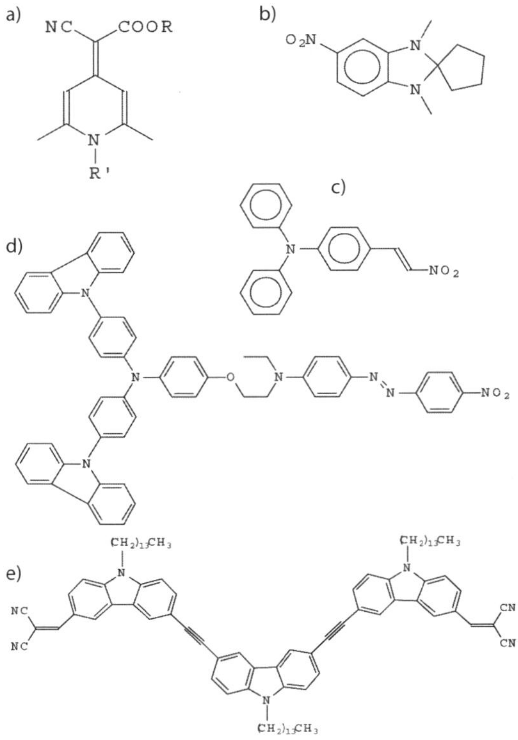

Fig. 24a–e. Examples of bifunctional chromophores for photorefractive applications: **a** N-2-butyl-2,6-dimethyl-4H-pyridone-4-ylidenecyanomethylacetate (2BNCM); **b** 1,3-dimethyl1-2,2-tetramethylene-5-nitrobenzimidazoline (DTNBI); **c** 4-(N,N'-diphenylamino)-(β)-nitrostyrene (DPANST); **d** 4,4'-di(N-carbazolyl)-4''-(2-N-ethyl-4-[2-(4-nitrophenyl)-1-azo]anilinoethoxy)-triphenylamine (DRDCTA); **e** carbazole trimer

glass transition temperature and form films at room temperature with good optical quality. Mixtures of a bifunctional carbazole trimer and TNF as a sensitizer showed a net gain of 76 cm^{-1} at an applied field of 30 V/μm.

If bifunctional molecules do not form glasses, they can be incorporated into polymer binders. Examples of bifunctional molecules are shown in Fig. 24. DTNBI (molecule (b) in Fig. 24) was doped in PMMA and C$_{60}$ was used as a sensitizer. Diffraction efficiencies of 7%, sub-second grating growth times, and net two-beam coupling gain coefficients of 34 cm^{-1} were observed in such samples [81]. DPANST (molecule (c) in Fig. 24) was doped into PBMA [84]. Recently, ms response times were reported in photorefractive glasses based on the bifunctional chromophore DRDCTA (molecule (d) in Fig. 24) doped with the plasticizer DOP (molecule (b) in Fig. 23) and C$_{60}$ as a sensitizer [89, 90].

Another approach to provide materials with long-term stability is to synthesize fully functionalized polymers. This route is more challenging than any of the guest/host approaches described so far and requires a major effort in chemical synthesis. It offers less flexibility than the guest-host approach. The performance of fully functionalized polymers has not yet reached the performance level of guest/host systems; however, several interesting materials have been proposed. Several materials including conjugated polymers were developed [91–94]. Other approaches are based on the combination of organic and inorganic materials such as in sol-gel approaches [95–96] and the use of inorganic semiconductor quantum dots as sensitizers [97].

In the amorphous organic photorefractive materials discussed above, the electro-active molecules are assumed to be non-interacting. Their orientation in an electric field requires consequently high field values (typically a few tens of V/μm). In contrast, liquid crystals exhibit electroactivity at much lower fields. Photorefractivity in liquid crystals was first demonstrated by Rudenko and Sukhov in the nematic liquid crystal 5CB (see molecule (h) in Fig. 19) doped with the dye rhodamine 6G [98]. Under excitation of an argon ion laser, the dye dissolved in the liquid crystal undergoes reversible heterolytic dissociation and leads to the formation of ions. The drift and diffusion of these ions in the mixture lead to a space-charge field that reorients the axis (or director) of the liquid crystal. Since this orientational effect is a Kerr effect and is quadratic in electric field, a d.c. field is applied to the sample. In contrast to photorefractive polymers for which the applied voltage is several kilovolts, here the applied d.c. voltage required for efficient wave-mixing effects is approximately 1 V for a 100 μm-thick film [99]. High photorefractive gain was observed in nematic liquid crystal mixtures of 5CB and 8OCB doped with electron donor and acceptor molecules [100]. In this case, mobile ions were produced through photoinduced electron transfer reactions between the acceptor N,N'-di(n-octyl)-1,4,5,8-naphthalenediimide and the donor perylene. Due to the large coherent length of a liquid crystalline phase, the resolution of these materials is limited. Therefore, most of the wave-mixing experiments were performed with writing beams intersecting in the sample with a small angle, leading to large values of the grating spacing. Consequently, most two-beam coupling experiments were performed in the thin grating limit and the data were analyzed with coupled-wave theory for thick gratings leading to large values of gain coefficients. These values cannot be compared with those measured in polymers in the thick grating limit. To improve the resolution of photorefractive liquid crystals, polymer stabilized nematic liquid crystals were designed [101]. In this case, the addition of low concentrations of a polymeric electron acceptor creates an anisotropic gel-like medium in which higher resolution is achieved. Unfortunately, this network does also slow down the ionic transport and increases the response time constant. Since transport of charge in photorefractive liquid crystals is achieved through ionic transport rather than intermolecular electronic hopping processes,

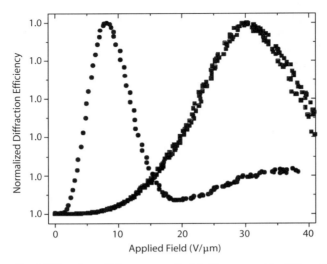

Fig. 25. Normalized diffraction efficiency vs applied field measured in 105 μm-thick samples of the polymer-dispersed liquid crystal TL202:PMMA:ECZ:TNFDM (*circles*) and the photorefractive polymer DHADC-MPN:PVK:ECZ:TNF (*squares*)

charge mobility is limited. Another approach that combines high speed, high resolution, and low field reorientation of a liquid crystalline phase is to use polymer dispersed liquid crystals. In such materials the space-charge field is written in a photoconducting polymer matrix as in a traditional photorefractive polymer. This space-charge field reorients the director of the liquid crystal droplets dispersed in the polymer matrix and leads to an effective refractive index modulation [102, 103]. In 105 μm-thick films of PMMA doped with ECZ as a transport material, TNFDM as a sensitizer, and the liquid crystal mixture TL202 purchased from Merck, overmodulation of the diffraction efficiency at a grating spacing of 3 μm was observed at an applied field of 8 V/μm as shown in Fig. 25 [104].

7
Applications

Optical media that can record elementary gratings, such as photorefractive polymers, are in high demand. In many fields, the decomposition of a signal into a superposition of harmonic functions (Fourier analysis) is a powerful tool. In optics, Fourier analysis often provides an efficient way to implement complex operations and is at the basis of many optical systems. For instance, any arbitrary image can be decomposed into a sum of harmonic functions with different spatial frequencies and complex amplitudes. Each of these periodic functions can be considered as an elementary grating. For real-time optical recording or processing applications, the material must have a dynamic response. In other words, the light-induced gratings should be erasable or be able to accommodate any changes in the light waves that are inducing them, in real time. Materials with such optical encoding properties allow for implementation of a wide range of optical applications ranging from reconfigurable interconnects, dynamic holographic storage, to optical correlation, image recognition, image processing, and phase conjugation. Therefore, materials where the optical encoding is dynamic and based on a periodic modulation of the refractive index are the focus of intense research. In this section we will provide some examples of applications that have been demonstrated so far with photorefractive polymers. As this new class of materials become more mature, a plethora of other applications are anticipated for the future.

Holographic storage has been a strong driver during the early development of photorefractive polymers. Early demonstration of hologram recording and retrieval was performed in DMNPAA:PVK:ECZ:TNF samples and was made possible by the high diffraction efficiency measured in 105 µm-thick samples [105]. More detailed assessments of their potential for holographic storage were conducted with the holographic storage test-bed developed at IBM. In samples of the composite 2BNCM:PMMA:TNF, single digital data pages could be stored and retrieved using only 1 part in 10^6 of the total dynamic range [87]. High-contrast 64-kbit digital data pages could be stored in 130 µm-thick DTNBI:PMMA:C_{60} samples placed near the Fourier plane in a *4-f* recording geometry. Due to the limited thickness of the sample, only single data pages were stored. A global absolute threshold resulted in a readout BER (Bit Error Rate) of 1.5×10^{-4} [106]. By using polysiloxane polymers substituted with carbazole side-groups and doped with FDEANST and TNF, the IBM group demonstrated digital data recording at a density of 0.52 Mbit/cm^2 [107].

As a reconfigurable recording medium, photorefractive polymers have also been used for dynamic time-average interferometry [105], wave-front phase conjugation and phase doubling [108], and incoherent-to-coherent image conversion [108]. To demonstrate the technological potential of high efficiency photorefractive polymers in device geometry, an all-optical, all-polymeric pattern recognition system for security verification was demonstrated [109]. In this system, the photorefractive polymer was used as the real-time optical recording and processing medium. In the proposed security verification system, documents were optically encoded with pseudo-randomly generated phase masks and they were inspected by performing all-optical spatial correlation of two phase encoded images in a photorefractive polymer. To authenticate the document it was compared with a master, which was an exact copy of the mask. The hologram written by the interference of a reference beam and a laser beam going through the test mask formed a holographic filter for the master mask. If the two phase patterns matched, light was strongly diffracted by the photorefractive polymer. The detection of that light

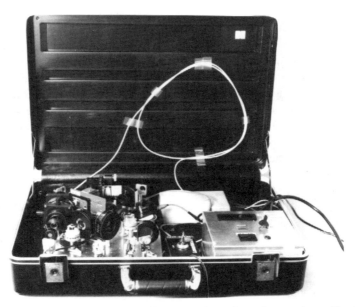

Fig. 26. Photograph of a compact version of the optical correlator for security verification that uses photorefractive polymers

provided a way to check the authenticity of the document that was tested. This system has many advantages: the use of a highly efficient photorefractive polymer as active material and its compatibility with semiconductor laser diodes keep the overall manufacturing cost to levels that are significantly lower than that of any previous proposed optical correlator. The system is fast because the processing is implemented optically in parallel. Furthermore, the high resolution of the photorefractive polymers allows the use of shorter focal length lenses in the 4-f correlator, thus making its design more compact compared with one using liquid crystal light valves. Figure 26 shows a photograph of a compact version of the system for security verification that was built.

While the previous application was based on real-time four-wave mixing effects, other applications rely on the strong coherent two-beam coupling effects observed in photorefractive polymers. Such applications include self-pumped phase conjugation mirrors [110]. In this case, high optical gain is required and was obtained by stacking several photorefractive polymer samples of PD-CST:PVK:BBP:C_{60}. A single pass optical gain of 5 could be demonstrated as shown in Fig. 27. Beam-coupling in photorefractive polymers can also be used for optical limiting [111]. In this case, the application relies on the existence of amplified scattering, an effect referred to as beam fanning [112, 113].

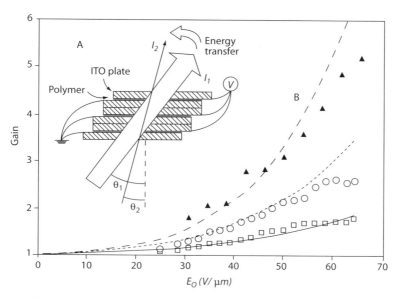

Fig. 27. Field dependence of the two-beam coupling gain measured in the photorefractive polymer composite PDCST:PVK:BBP:C_{60}; *squares*: single sample (140 μm-thick); *circles*: a two-layer stack; *triangles*: a three-layer stack

The real-time holographic recording properties of photorefractive polymers have also been exploited in medical imaging applications by performing time-gated holography [80, 114]. In this method a hologram is formed by the temporal overlap in the photorefractive polymer of the reference pulse and the first-arriving (least-scattered) light from the stretched image-bearing pulse. The filtering of the useful photons from the scattered photons is thus achieved in real time without any need for digital processing. The image carried by the ballistic light can then be reconstructed in real time via diffraction of a probe beam (pulsed or continuous-wave). An example of a reconstructed image using this technique is shown in Fig. 28. The imaging was performed in a degenerate four-wave mixing geometry using a femtosecond Ti:Sapp laser at 800 nm in the near infra-red. The object was an Air Force resolution target positioned behind a 10 mm-thick cell filled with a water suspension of polystyrene micro-spheres serving as a model scattering medium. The image was obtained after the laser beam carrying the information was passing through a 0.06% suspension of 0.5-μm microspheres, which is approximately nine mean free paths, or an effective optical density of 4. Performance achievable with photorefractive polymers was comparable to or exceeded what is typically achieved with photorefractive inorganic crystals. In addition, due to their low cost and their ability to be processed into large area films, photorefractive polymers have greater technological potential than their inorganic counterparts.

Fig. 28. Portion of an Air Force bar-target imaged through nine scattering mean free paths using low-coherence holographic time-gating in a photorefractive polymer in the near infra-red (800 nm)

Other applications that were recently demonstrated with photorefractive polymers include homodyne detection of ultrasonic surface displacements using two-wave mixing [115]. With the development of new photorefractive polymers with response times in the millisecond range, numerous optical processing techniques can be performed at video rates. Image amplification and novelty filtering at video rates were demonstrated recently [116]. All-optical processing techniques compete with computational methods. Therefore, it is important that photorefractive polymers exhibit faster response times in the future.

8
Conclusion and Outlook

Despite the high efficiency of existing materials and their response times that enable applications at video rates, much effort is still needed in the field before this class of optical materials reaches its full potential. A lot of the early advances were made through the combination of different materials and did establish that photorefractive polymers were suitable for various real-time holographic applications that use low cost low power laser diodes. However, a rational design that leads to controlled properties is required to push these materials to their ultimate performance level. While design guidelines to optimize the refractive index change have been proposed and demonstrated experimentally, very little is known about the photoconducting properties of these multifunctional materials. Indeed, the origin of the photorefractive trap in materials doped with the sensitizer C_{60} has only been discussed recently [69] and illustrates the need for further studies to develop a better understanding of the physical and chemical properties of these complex materials. Owing to the progress that is made in the field of organic optical materials in general and the development of structure-property relationships in organic molecules and polymers, the future of photorefractive polymers looks bright. The ability to control the relative energy levels of the frontier orbitals of the different molecules that are doped into these polymer composites [70], as well as the electron transfer rates between these levels, will play an important role in the future. Current materials are amorphous or consist of nematic liquid crystalline phases. The development of new materials including supramolecular structures through nanofabrication technologies or hybrid materials that combine organic and inorganic materials will bring photorefractive polymers to new performance levels. Materials with microsecond response times will emerge with sensitivities extended to the telecommunication wavelengths. New applications for these materials will be generated as their performance, modeling, and manufacturing further advance.

Acknowledgements. This work was supported by NSF through an ECS research grant and through a CAREER grant, by the AFOSR, and ONR through the MURI Center CAMP. The authors would like to acknowledge Prof. S. R. Marder for his collaboration, for introducing us to the bond length alternation theory to describe the linear and nonlinear optical properties of organic molecules, and for providing us with numerous organic molecules and polymers for our work.

References

1. NRC (1998) Harnessing light: optical science and engineering for the 21st century. National Research Council, Committee on Optical Science and Engineering. National Academy Press, Washington
2. Moerner WE, Grunnet-Jepsen A, Thompson CL (1997) Annu Rev Mater Sci 27:585
3. Meerholz K, Kippelen B, Peyghambarian N (1998) In: Wise DL, Wnek GE, Trantolo DJ, Cooper TM, Gresser JD (eds) Photonic polymer systems. Marcel Dekker, New York
4. Kippelen B, Peyghambarian N (1997) In: Andrews MP, Najafi SI (eds) Sol-gel and polymer photonic devices, critical review of optical science and technology. CR 68. SPIE Optical Engineering Press, Bellingham
5. Arfken G (1985) Mathematical methods for physicists. Academic Press, Boston
6. Born M, Wolf E (1980) Principles of optics. Pergamon Press, Oxford
7. Franken PA, Hill AE, Peters CW, Weinreich G (1961) Phys Rev Lett 7:118
8. Bloembergen N (1964) Nonlinear optics. Benjamin, New York
9. Shen YR (1984) The principles of nonlinear optics. Wiley, New York
10. Boyd RW (1992) Nonlinear optics. Academic Press, Boston
11. Yariv A (1975) Quantum electronics. Wiley, New York
12. Oudar JL, Chemla DS (1975) Opt Commun 13:164
13. Chemla DS, Zyss J (1987) Nonlinear optical properties of organic molecules and crystals. Academic Press, Orlando
14. Zyss J (1994) Molecular nonlinear optics: materials, physics and devices. Academic Press, Boston
15. Kanis DR, Ratner MA (1994) Chem Rev 94:195
16. Marder SR, Beratan, DN, Cheng LT (1991) Science 252:103
17. Gorman CB, Marder SR (1993) Proc Natl Acad Sci USA 90:11,297
18. Meyers F, Marder SR, Pierce BM, Brédas JL (1994) J Am Chem Soc 116:10,703
19. Marder SR, Kippelen B, Jen AKJ, Peyghambarian N (1997) Nature 388:845

20. Williams DJ (1987) In: Chemla DS, Zyss J (eds) Nonlinear optical properties of organic molecules and crystals. Academic Press, New York
21. Singer KD, Kuzyk MG, Sohn JE (1987) J Opt Soc Am B 4:968
22. Burland DM, Miller RD, Walsh CA (1994) Chem Rev 94:31
23. Yu YZ, Wong KY, Garito AF (1997) In: Nalwa HS, Miyata S (eds) Nonlinear optics of organic molecules and polymers. CRC Press, Boca Raton
24. Wu JW (1991) J Opt Soc Am B 8:142
25. Ashkin A, Boyd GD, Dziedzic JM, Smith RG, Ballmann AA, Nassau K (1966) Appl Phys Lett 9:72
26. Chen FS (1967) J Appl Phys 38:3418
27. Chen FS (1969) J Appl Phys 40:3389
28. Amodei JJ (1971) Appl Phys Lett 18:22
29. Günter P (1982) Phys Rep 93:199
30. Feinberg J (1983) In: Fisher RA (ed) Optical phase conjugation. Academic Press, New York
31. Günter P, Huignard JP (1988) (1989) Photorefractive materials and their applications, vols I and II. Springer, Berlin Heidelberg New York
32. Nolte DD (1995) Photorefractive effects and materials. Kluwer, Boston
33. Roosen G (1989) Int J Optoelectron 4:459
34. Nolte DD, Olson DH, Doran GE, Knox WH, Glass AM (1990) J Opt Soc Am B 7:2217
35. Wang Q, Nolte DD, Melloch MR (1991) Appl Phys Lett 59:256
36. Sutter K, Günter P (1990) J Opt Soc Am B 7:2274
37. Ducharme S, Scott JC, Twieg RJ, Moerner WE (1991) Phys Rev Lett 66:1846
38. Kukhtarev NV, Markov VB, Soskin M, Vinetskii VL (1979) Ferroelectrics 22:949
39. Yeh P (1993) Introduction to photorefractive nonlinear optics. Wiley, New York
40. Schildkraut JS, Cui YJ (1992) J Appl Phys 72:5055
41. Eichler HJ, Günter P, Pohl DW (1986) Laser-induced dynamic gratings. Springer, Berlin Heidelberg New York
42. Kogelnik H (1969) Bell Syst Tech J 48:2909
43. Moerner WE, Silence SM, Hache F, Bjorklund GC (1994) J Opt Soc Am B 11:320
44. Schildkraut JS (1990) Appl Phys Lett 58:340
45. Tamura K, Padias AB, Hall HK Jr, Peyghambarian N (1992) Appl Phys Lett 60:1803
46. Yu L, Chan W, Bao Z, Cao SXF (1993) Macromolecules 26:2216
47. Donckers MCJM, Silence SM, Walsh CA, Scott JC, Matray TJ, Twieg RJ, Hache F, Bjorklund GC, Moerner WE (1993) Opt Lett 18:1044
48. Kippelen B, Sandalphon, Peyghambarian N, Lyon SR, Padias AB, Hall HK Jr (1993) Electron Lett 29:1873
49. Meerholz K, Volodin B, Sandalphon, Kippelen B, Peyghambarian N (1994) Nature 371:497
50. Cox AM, Blackburn RD, West DP, King TA, Wade FA, Leigh DA (1996) Appl Phys Lett 68:2801
51. Hendrickx E, Volodin BL, Steele DD, Maldonado JL, Wang JF, Kippelen B, Peyghambarian N (1997) Appl Phys Lett 71:1159
52. Hendrickx E, Herlocker J, Maldonado JL, Marder SR, Kippelen B, Persoons A, Peyghambarian N (1998) Appl Phys Lett 72:1679
53. Law KY (1993) Chem Rev 93:449
54. Borsenberger PM, Weiss DS (1993) Organic photoreceptors for imaging systems. Marcel Dekker, New York
55. Onsager L (1938) Phys Rev 54:554
56. Mozumder A (1974) J Chem Phys 60:4300
57. Mort J (1980) Adv Phys 29:367
58. Scher H, Montroll EW (1975) Phys Rev B 12:2455
59. Schmidlin SW (1977) Phys Rev B 16:2362

60. Bässler H (1993) Adv Mat 5:662
61. Borsenberger PM, Magin EH, Van der Auweraer M, De Schryver FC (1993) Phys Stat Sol (a) 140:9
62. Borsenberger PM, Bässler H (1994) J Appl Phys 75:967
63. Borsenberger PM, Detty MR, Magin EH (1994) Phys Stat Sol (b) 185:465
64. Borsenberger PM, Magin EH, Van der Auweraer M, De Schryver FC (1994) Phys Stat Sol (b) 186:217
65. Borsenberger PM, Gruenbaum WT, Magin EH (1995) Phys Stat Sol (b) 190:555
66. Borsenberger PM, Shi J (1995) Phys Stat Sol (b) 191:461
67. Borsenberger PM, Gruenbaum WT, Sorriero LJ, Zumbulyadis N (1995) Jpn J Appl Phys 34:L1597
68. Borsenberger PM, Magin EH, O-Regan MB, Sinicropi JA (1996) J Polym Sci B Polym Phys 34:317
69. Grunnet-Jepsen A, Wright D, Smith B, Bratcher MS, DeClue MS, Siegel JS, Moerner WE (1998) Chem Phys Lett 291:553
70. Herlocker JA, Fuentes-Hernandez C, Ferrio KB, Hendrickx E, Zhang Y, Wang JF, Marder SR, Blanche PA, Peyghambarian N, Kippelen B (2000) Appl Phys Lett (in press)
71. Wortmann R, Poga C, Twieg RJ, Geletneky C, Moylan CR, Lundquist PM, DeVoe RG, Cotts PM, Horn H, Rice JE, Burland DM (1996) J Chem Phys 105:10,637
72. Kippelen B, Meyers F, Peyghambarian N, Marder SR (1997) J Am Chem Soc 119:4559
73. Kippelen B, Meerholz K, Peyghambarian N (1997) In: Nalwa HS, Miyata (eds) Nonlinear optics of organic molecules and polymers. CRC Press, Boca Raton
74. Zhang Y, Burzynski R, Ghosal S, Casstevens MK (1996) Adv Mater 8:111
75. Meerholz K, De Nardin Y, Bittner R, Wortmann R, Würthner F (1998) Appl Phys Lett 73:4
76. Wright D, Díaz-García MA, Casperson JD, DeClue M, Moerner WE, Twieg RJ (1998) Appl Phys Lett 73:1490
77. Herlocker JA, Ferrio KB, Guenther BD, Mery S, Kippelen B, Peyghambarian N (1999) Appl Phys Lett 74:2253
78. Zhang J, Singer KD (1998) Appl Phys Lett 72:2948
79. Hendrickx E, Kippelen B, Thayumanavan S, Marder SR, Persoons A, Peyghambarian N (2000) J Chem Phys 112:9557
80. Kippelen B, Marder SR, Hendrickx E, Maldonado JL, Guillemet G, Volodin B, Steele DD, Enami Y, Sandalphon, Yao YJ, Wang JF, Röckel H, Erskine L, Peyghambarian N (1998) Science 279:54
81. Silence SM, Scott JC, Stankus JJ, Moerner WE, Moylan CR, Bjorklund GC, Twieg RJ (1995) J Phys Chem 99:4096
82. Wortmann R, Poga C, Twieg RJ, Geletneky C, Moylan CR, Lundquist PM, DeVoe RG, Cotts PM, Horn H, Rice JE, Burland DM (1996) J Chem Phys 105:10,637
83. Burzynski R, Zhang Y, Ghosal S, Casstevens MK (1995) J Appl Phys 78:6903
84. Zhang Y, Ghosal S, Casstevens MK, Burzynski R (1995) Appl Phys Lett 66:256
85. Okamoto K, Nomura T, Park SH, Ogino K, Sato H (1999) Chem Mater 11:3279
86. Grunnet-Jepsen A, Thompson CL, Twieg RJ, Moerner WE (1997) Appl Phys Lett 70:1515
87. Lundquist PM, Wortmann R, Geletneky C, Twieg RJ, Jurich M, Lee VY, Moylan CR, Burland DM (1996) Science 274:1182
88. Wang L, Zhang Y, Wada T, Sasabe H (1996) Appl Phys Lett 69:728
89. Schloter S, Schreiber A, Grasruck M, Leopold A, Kol'chenko M, Pan J, Hohle C, Strohriegl P, Zilker SJ, Haarer D (1999) Appl Phys B 68:899
90. Hohle C, Hofmann U, Schloter S, Thelakkat M, Strohriegl P, Haarer D, Zilker SJ (1999) J Mater Chem 9:2205
91. Kippelen B, Tamura K, Peyghambarian N, Padias AB, Hall HK Jr (1993) Phys Rev B 48:10,710
92. Zhao C, Park CK, Prasad PN, Zhang Y, Ghosal S, Burzynski R (1995) Chem Mater 7:1237

93. Yu L, Chen YM, Chan WK (1995) J Phys Chem 99:2797
94. Li L, Chittibabu KG, Chen Z, Chen JI, Marturunkakul S, Kumar J, Tripathy SK (1996) Opt Commun 125:257
95. Burzynski R, Casstevens MK, Zhang Y, Ghosal S (1996) Opt Eng 35:443
96. Chaput F, Riehl D, Boilot JP, Cargnelli K, Canva M, Levy Y, Brun A (1996) Chem Mater 8:312
97. Winiarz JG, Zhang L, Lal M, Friend CS, Prasad PN (1999) J Am Chem Soc 121:5287
98. Rudenko EV, Sukhov AV (1994) JETP Lett 59:142
99. Khoo IC, Li H, Liang Y (1994) Opt Lett 19:1723
100. Wiederrecht GP, Yoon BA, Wasielewski MR (1995) Science 270:1794
101. Wiederrecht GP, Wasielewski MR (1998) J Am Chem Soc 120:3231
102. Golemme A, Volodin BL, Kippelen B, Peyghambarian N (1997) Opt Lett 22:1226
103. Ono H, Kawatsuki N (1997) Opt Lett 22:1144
104. Golemme A, Kippelen B, Peyghambarian N (1998) Appl Phys Lett 73:2408
105. Volodin BL, Sandalphon, Kippelen B, Kukhtarev NV, Peyghambarian N (1995) Opt Eng 34:2213
106. Poga C, Lundquist PM, Lee V, Shelby RM, Twieg RJ, Burland DM (1996) Appl Phys Lett 69:1047
107. Lundquist PM, Poga C, DeVoe RG, Jia Y, Moerner WE, Bernal MP, Coufal H, Grygier RK, Hoffnagle JA, Jefferson CM, Macfarlane RM, Shelby RM, Sincerbox GT (1996) Opt Lett 21:890
108. Volodin BL, Kippelen B, Meerholz K, Peyghambarian N, Kukhtarev NV, Caulfield HJ (1996) J Opt Soc Am B 13:2261
109. Volodin BL, Kippelen B, Meerholz K, Javidi B, Peyghambarian N (1996) Nature 383:58
110. Grunnet-Jepsen A, Thompson CL, Moerner WE (1997) Science 277:549
111. Grunnet-Jepsen A, Thompson CL, Moerner WE (1997) Mat Res Soc Symp Proc 479:199
112. Grunnet-Jepsen A, Thompson CL, Twieg RJ, Moerner WE (1998) J Opt Soc Am B 15:901
113. Meerholz K, Bittner R, De Nardin Y (1998) Opt Commun 150:205
114. Steele DD, Volodin BL, Savina O, Kippelen B, Peyghambarian N (1998) Opt Lett 23:153
115. Klein MB, Bacher GD, Grunnet-Jepsen A, Wright D, Moerner WE (1999) SPIE 3589:22
116. Goonesekera A, Wright D, Moerner WE (2000) Appl Phys Lett 76:3358

Received June 2001

Organics and Polymers with High Two-Photon Activities and their Applications

Tzu-Chau Lin · Sung-Jae Chung · Kyoung-Soo Kim · Xiaopeng Wang
Guang S. He · Jacek Swiatkiewicz · Haridas E. Pudavar · Paras N. Prasad*

Departments of Chemistry and Physics, Institute for Lasers, Photonics and Biophotonics, State University of New York at Buffalo, Buffalo, New York 14260-3000, USA
* E-mail: pnprasad@acsu.buffalo.edu

This review describes some of the recent developments in materials which exhibit enhanced two-photon absorption that can initiate photopolymerization or up-converted emission. Various optical methods including femtosecond time-resolved pump-probe experiments to characterize the two-photon properties are discussed. Finally, the applications of two-photon processes to optical power limiting, up-converted lasing, 3-D data storage, 3-D micro-fabrication, two-photon fluorescence microscopy and bio-imaging, and two-photon photodynamic therapy are presented.

Keywords. Two-photon absorption, Non-linear transmission, Z-scan, Optical power limiting, Up-converted lasing

List of Symbols and Abbreviations 158

1   Introduction 159

2   Theoretical Background 160

3   Strategy for Molecular Design 162

4   Characterization 168

4.1   Non-Linear Transmission 168
4.2   Up-Converted Fluorescence Emission. 169
4.3   Transient Absorption 170
4.4   Four-Wave Mixing 172
4.5   Z-Scan Technique. 172

5   Applications 174

5.1   Optical Power Limiting. 174
5.2   Up-Converted Lasing. 176
5.3   3-D Data Storage 178

Advances in Polymer Science, Vol. 161
© Springer-Verlag Berlin Heidelberg 2003

5.4   3-D Microfabrication 180
5.5   Two-Photon Fluorescence Microscopy 181
5.6   Two-Photon Photodynamic Therapy 184

6   **Polymeric Materials** 185

7   **Summary** 189

References 190

List of Symbols and Abbreviations

μ_{ind}	dipole moment
e	electronic charge
r	induced displacement
P_{ind}	induced polarization vector
N	electron density
E	external electric field vector
ε	dielectric constant
$\chi^{(n)}$	optical susceptibility
ω	optical frequency
n_c	complex refractive index
n	real part of refractive index
k	imaginary part of refractive index
I	intensity of light
c	speed of light in vacuum ($3\times10^8\,\mathrm{ms^{-1}}$)
n_{photon}	the number of photons absorbed
N_m	the number of absorbing molecules per unit volume
F	photon flux of the light source
D	electron donor
A	electron acceptor
$\lambda_{max}(\mathrm{Abs})$	linear maximum absorption wavelength (nm)
$\lambda_{max}(\mathrm{Em})$	maximum emission wavelength (nm)
TPA	two-photon absorption
α	attenuation coefficient
T, T_o, T_I	optical transmissivity of a medium
L	sample thickness (m)
β	non-linear absorption coefficient
σ_2	molecular two-photon absorption cross-section ($\mathrm{cm^4\,GW^{-1}}$)
σ_2'	molecular two-photon absorption cross-section ($\mathrm{cm^4\,s\,photon^{-1}}$)
N_A	Avogadro constant (6.026×10^{23})

h	Plank constant (6.626×10^{-34} J s)
ν	frequency of light (s^{-1})
d_o	concentration (mol L^{-1})

1
Introduction

Extensive research has been conducted in the field of multi-photon spectroscopy for the past several decades. However, until recently, multi-photon processes did not find widespread applications due to the small multi-photon absorptivity of materials. The contributions from several research groups to develop a new generation of multifunctional organic materials with sufficiently large multi-photon absorption cross-sections have opened up a number of novel applications in photonics and biophotonics.

Multi-photon absorption is the process that occurs through simultaneous absorption of photons via virtual states in a medium. Although it was theoretically predicted by Maria Göppert-Mayer in 1931 [1], the requirement of a high peak-power laser made the first experimental evidence of this non-linear phenomenon come 30 years later when Kaiser and Garret performed two-photon excitation in a $CaF_2:Eu^{2+}$ crystal [2]. Some early potential applications were demonstrated by Rentzepis in data storage [3] and by Webb in microscopy [4]. However, due to the relatively small two-photon absorption cross-sections of most materials, it found limited usage. Over the past few years, this non-linear effect in organic systems has drawn a great deal of attention [5, 6] because organic materials provide the flexibility to modify their structures at the molecular level and to optimize their non-linear optical response. Reports of new organic chromophores with enhanced two-photon absorptivities and up-converted fluorescence provided by our research group [7–12] and other groups [13–15] have initiated considerable interest in the development of highly efficient two-photon materials and opened up a myriad of new applications including two-photon up-conversion lasing [16–20], two-photon optical power limiting [7–9, 21–23], 3-D optical data storage [3, 24–28], two-photon excited fluorescence for non-destructive bio-imaging [29–35], and two-photon photodynamic therapy [36]. In this review article, we present recent significant achievements in the improved two-photon absorptivity and report some molecular designing parameters. The fundamental theories of the two-photon absorption process and the experimental probes to investigate this optical response are briefly discussed. Finally, some of the recent applications utilizing two-photon absorption are also presented.

2
Theoretical Background

The interaction of an organic medium with the external electrical or optical field can be described within the framework of a dielectric subjected to an electric field. The induced dipole moment in a molecule due to the applied field is given by:

$$\mu_{ind} \equiv -e\boldsymbol{r} \tag{1}$$

where e is the electronic charge and r is the induced displacement due to the field. For the bulk medium, the polarization resulting from this induced dipole moment is given by:

$$\boldsymbol{P}_{ind} = -Ne\boldsymbol{r} \tag{2}$$

where N is the electron density in the medium. In the cases with relatively low external field strength, the relationship between the induced polarization $\boldsymbol{P}_{ind}(E)$ in a medium and the applied field $\boldsymbol{E}$ is linear and can be expressed as:

$$\boldsymbol{P}_{ind}(E) = \chi^{(1)}\boldsymbol{E} \tag{3}$$

where $\chi^{(1)}$ is the linear susceptibility and is related to the dielectric constant ε as:

$$\varepsilon = 1 + 4\pi\chi^{(1)} \tag{4}$$

It should be mentioned here that the term $\chi^{(1)}$ is a second-rank tensor and has nine components because it is related to all the components of the polarization vectors and electric field vectors. Therefore, the dielectric constant is also a tensor of rank 2. The optical response of a medium at an optical frequency ω can be represented equivalently by the complex refractive index n_c as:

$$n^2_c(\omega) = \varepsilon(\omega) = 1 + 4\pi\chi^{(1)}\omega \tag{5}$$

The complex refractive index can also be expressed as the sum of the real and the imaginary parts as:

$$n_c = n + ik \tag{6}$$

where the real part n corresponds to the dispersion of refractive index and the imaginary part k corresponds to linear absorption. If the external field is an intense electric field such as that generated by a high-power laser, the relationship between the polarization vector of the medium $\boldsymbol{P}_{ind}$ and the electric field vector $\boldsymbol{E}$ will no longer remain linear because, in this case, higher-order terms in the polarization can be quite significant. Therefore, Eq. (3) assumes the form:

$$\boldsymbol{P}_{(E)} = \chi^{(1)}\boldsymbol{E} + \chi^{(2)}\boldsymbol{EE} + \chi^{(3)}\boldsymbol{EEE} + \chi^{(4)}\boldsymbol{EEEE} + \dots \tag{7}$$

where $\chi^{(1)}$ is the linear susceptibility of the medium, $\chi^{(2)}$, $\chi^{(3)}$, $\chi^{(n)}$ are the second-, third- and n-th order non-linear susceptibilities, respectively.

Molecular systems can interact with optical fields in two possible ways, one is through dissipative processes and the other one is through parametric processes. In parametric processes, there is an energy-momentum exchange between different modes of the optical field but there is no energy exchange between the optical field and the molecules in the system. However, in dissipative processes, the energy is exchanged between the optical field and the molecules through absorption and emission. Therefore, non-linear dissipative processes are related to multi-photon absorption. As mentioned above, the imaginary part of non-linear susceptibility corresponds to energy transfer from the optical field to the medium and this energy transfer rate can be expressed as [37]:

$$\frac{dW}{dt} = <E \cdot P> \tag{8}$$

where E and P are the electric field and the polarization vectors, respectively. The brackets in this equation mean a time average over several cycles of the field. If we consider only dissipative processes in the expression for P, the first term involving $\chi^{(1)}$ describes a linear absorption. The even-order susceptibilities like $\chi^{(2)}$, $\chi^{(4)}$, etc, do not make a contribution to the dissipative processes except when the external field is a DC field. Therefore, the lowest-order non-linear absorption will be described by the imaginary part of $\chi^{(3)}$, which corresponds to two-photon absorption, (Stokes) Raman gain or reverse (anti-Stokes) Raman attenuation and, similarly, the imaginary part of $\chi^{(5)}$ relates to three-photon absorption. For monochromatic waves with frequency ω, Eq. (8) can be written as:

$$\frac{dW}{dt} = \frac{1}{2} \omega \, \text{Im} \, (E \cdot P) \tag{9}$$

For a degenerate two-photon absorption process (absorbing two photons of the same frequency), Eq. (9) can be transformed into the following expression [6]:

$$\frac{dW}{dt} = \frac{8\pi^2 \omega}{n^2 c^2} I^2 \, \text{Im}\left(\chi^{(3)}\right) \tag{10}$$

where I is the intensity of light and is defined as $I = EE^* \, nc/8\pi$. From the equation, it should be noted that the rate of energy absorption in this non-linear absorption process is not linearly but quadratically dependent on the light intensity. Instead of using the extinction coefficient for a linear absorption process, two-photon absorption is described in terms of a cross-section σ_2 (two-photon absorption cross-section, or TPA cross-section) as:

$$\frac{dn_{photon}}{dt} = \sigma_2 N F^2 \tag{11}$$

where dn_{photon}/dt is the number of photons absorbed per unit time, N is the number of absorbing molecules per unit volume, and $F = I/h\nu$ is the photon flux of the light source. Because $dW/dt = (dn_{photon}/dt)\, h\nu$, one can combine this relation with Eqs. (10) and (2), to obtain the following, commonly used theoretical expression for σ_2 [6]:

$$\sigma_2 = \frac{8\pi^2 h\nu^2}{n^2 c^2 N} \mathrm{Im}\left(\chi^{(3)}\right) \tag{12}$$

The detailed experimental determination and conventional units for every physical term of the two-photon absorption cross-section will be discussed later in the characterization section.

3
Strategy for Molecular Design

In order to realize the full potential of the two-photon technology, major improvements are necessary in the design and synthesis of highly active organic two-photon chromophores together with requirements of solubility and photostability for the practical purposes. To facilitate the design and synthesis of new, efficient dye molecules, an investigation is needed to establish well-defined structure-property relationships for a large number of organic structures with systematically varied molecular structural factors and precisely reproducible characterization of two-photon properties. Although a very detailed knowledge of molecular design parameters of new chromophores with enhanced two-photon absorption cross-sections has been lacking, several research groups over the past few years have attempted to define some design strategies. Here, we briefly collect and organize some of the molecular design concepts that have emerged from our collaboration with the Polymer Branch of the Materials Directorate at the Air Force Research Laboratory in Dayton. Some other relatively important structural motifs developed by other groups are also discussed.

The major design concept proposed by the U.S. Air Force Research Laboratory and our group [10] is based on the relationship between the molecular two-photon absorption cross-section and the imaginary component of the third-order, non-linear optical susceptibility, Eq. (12). Our initial study focused on two general organic structural types as illustrated in Fig. 1. Type I chromophores are symmetrical in nature consisting of a polarizable π-electron bridge in the middle, flanked with heterocyclic electron-deficient groups on either side. Type II chromophores are structurally asymmetrical molecules consisting of a highly fluorescent aromatic and/or olefinic bridge, end-capped by an electron-donor on one side and an electron-acceptor on the other side.

Based on this proposed design concept, a series of organic molecules was synthesized particularly for the imaging applications and studied at 800 nm, an optimum wavelength at which most organic materials and biological tissues have large

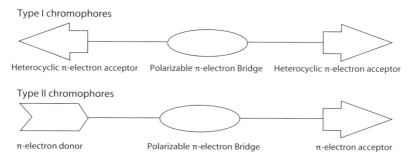

Fig. 1. Structural elements investigated for enhanced two-photon absorption

optical transparency. Also, widely available, mode-locked Ti-Sapphire lasers provide femtosecond pulses at this wavelength to produce efficient excitation. For this study, the challenge for the molecular design is to increase the molecular two-photon absorption cross-sections without shifting the two-photon absorption peak away from 800 nm. It was found that structural elements which can increase the effective conjugation length and polarizability of the molecule would enhance the molecular two-photon absorption cross-section. Fig. 2 presents the linear and non-linear optical properties of several representative chromophores synthesized by U.S. Air Force and investigated by our laboratory over the past few years. Some important structural factors/two-photon property relationships are also illustrated. The characterization methods and the units utilized for the molecular two-photon absorption cross-section (σ_2) are described in detail in the characterization section of this article. Two general criteria for increasing σ_2 at 800 nm can be derived from these studied organic structures:

1. Extending the π-conjugation length by incorporating more polarizable double bonds such as vinyl, heterocyclic, or aromatic moieties.
2. Increasing the planarity of the chromophore by utilizing fused aromatic rings as the π-bridge between the electron-donor and acceptor.

It should be noticed that the molecular design parameters described here are generally for the chromophores with their two-photon absorption peaks close to 800 nm because this is the wavelength of our interest for two-photon imaging applications. Furthermore, since the two-photon properties of these chromophores were measured by nanosecond pulses, it is possible that excited state absorption would enhance the reported σ_2 values. If one wishes to find out the absolute relationship between molecular structure and two-photon absorption cross-section of the studied molecules, the entire two-photon absorption spectra should be determined, although this measurement is not easy to conduct due to the lack of a high-power laser system providing a broad tunable range.

Type I Chromophores

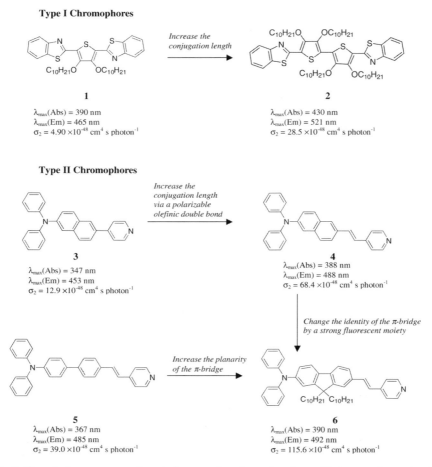

Type II Chromophores

Fig. 2. Chemical structures and optical properties of type I and type II chromophores developed by Reinhardt et al. From [10]

Recently, we have also applied some of the design strategies mentioned above into our chromophore systems by the combination of a polarizable stilbene as the π-bridge and a heterocyclic structure, oxadiazole, as the electron-acceptor. A new series of organic molecules with a multi-branched structure based on the linkage of two or three identical, asymmetric-type chromophore units to a common electron-donor group were synthesized [11]. Preliminary experimental results of their two-photon properties are promising and this molecular design may lead to new criteria for the development of multi-branched, multi-functional polymers and dendrimers with considerably enhanced two-photon absorptivities. The struc-

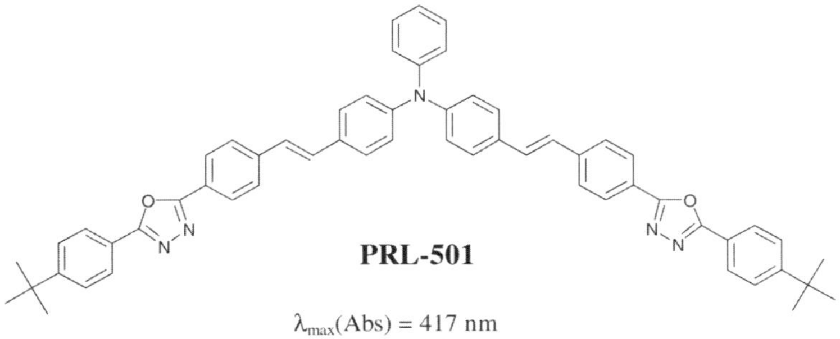

PRL-501

λ_{max}(Abs) = 417 nm
λ_{max}(Em) = 510 nm
$\sigma_2 = 208 \times 10^{-20}$ cm^4 GW^{-1}

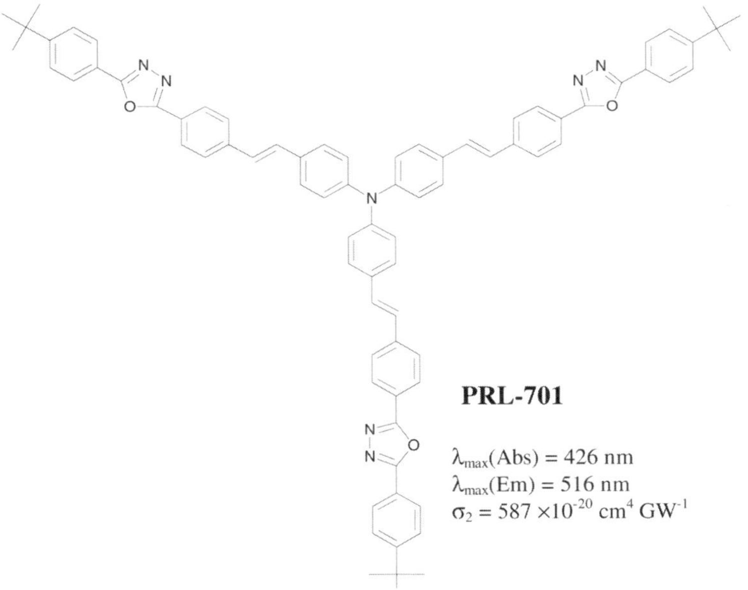

PRL-701

λ_{max}(Abs) = 426 nm
λ_{max}(Em) = 516 nm
$\sigma_2 = 587 \times 10^{-20}$ cm^4 GW^{-1}

Fig. 3. Chemical structures of multi-branched chromophores, **PRL-501** and **PRL-701**. From [11]

tures and optical properties of this newly synthesized multi-chromophore system are illustrated in Fig. 3. In collaboration with Professor Fréchet's group, we have also developed a dendritic system. These dendritic structures are functionalized with the analogue of our multi-branched chromophore (mentioned above) at the chain ends and we have studied their two-photon properties in that work [12].

Other research groups have also attempted to establish structure-property relationships by testing different organic functional groups as the building structures

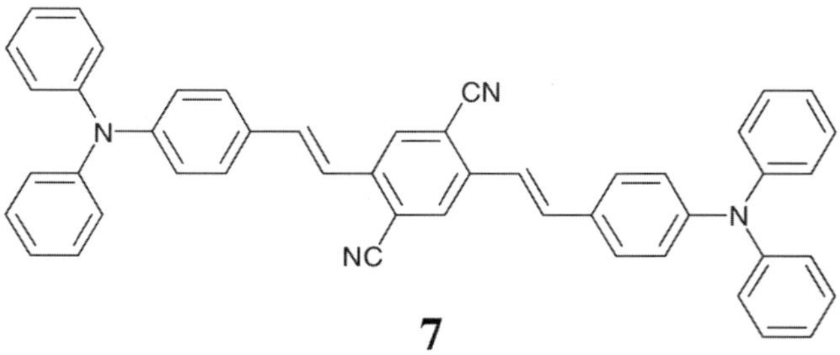

7

$\lambda_{max}(Abs) = 472$ nm
$\lambda_{max}(Em) = 525$ nm
$\sigma_2 = 19.4 \times 10^{-48}$ cm^4 s photon^{-1}

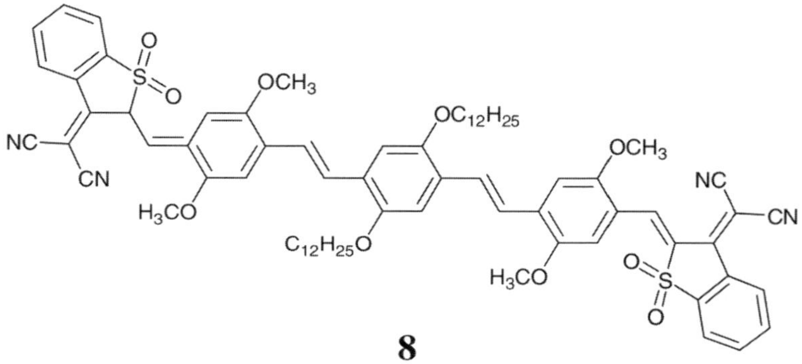

8

$\lambda_{max}(Abs) = 618$ nm
$\lambda_{max}(Em) = 745$ nm
$\sigma_2 = 44.0 \times 10^{-48}$ cm^4 s photon^{-1}

Fig. 4. Chemical structures and optical properties of D-A-D and A-D-A type chromophores (Marder and Perry et al.). From [14]

for two-photon chromophores. Marder, Perry and coworkers have studied the effect of electron-donor strength on stilbene derivatives [13, 14]. Chromophores designed and synthesized in their study are all structurally symmetrical in nature and can be categorized into two different types: D-A-D type and A-D-A type. Fig. 4 shows the structures and the measured σ_2 values of an example from each type as reported by them. They also reported that the peak position of two-photon absorption occurs at wavelengths shorter than two times the linear absorption λ_{max}, which means that the excited states achieved through two-photon process are higher than those achieved via linear absorption.

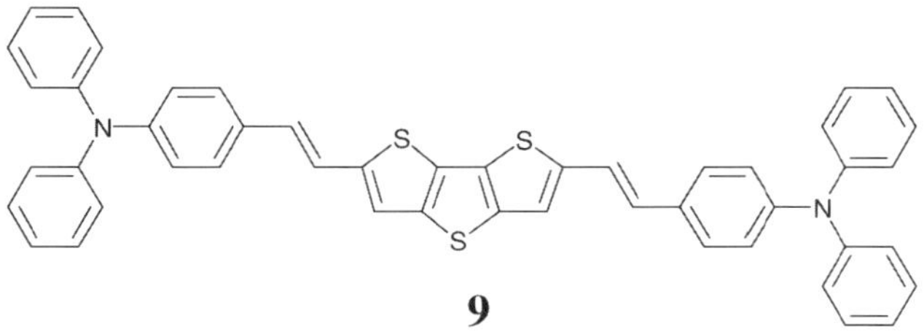

9

$\lambda_{max}(Abs) = 453$ nm (479 nm shoulder)
$\lambda_{max}(Em) = 512$ nm (543 nm shoulder)
$\sigma_2 = 810 \times 10^{-20}$ cm^4 GW^{-1}

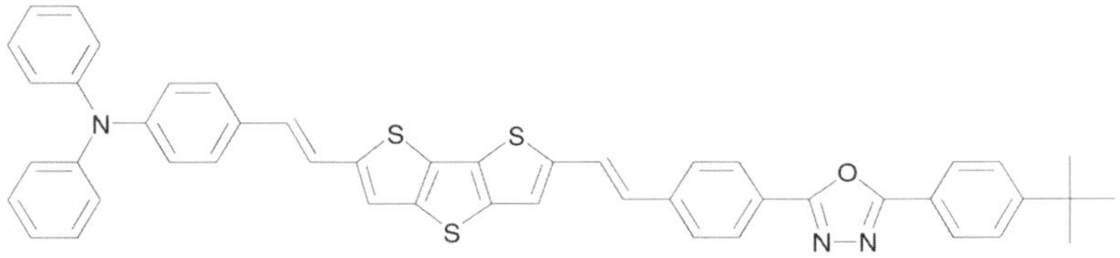

10

$\lambda_{max}(Abs) = 456$ nm
$\lambda_{max}(Em) = 565$ nm
$\sigma_2 = 483 \times 10^{-20}$ cm^4 GW^{-1}

Fig. 5. Chemical structures and optical properties of DDT-containing chromophores (Kim et al.). From [15]

Our collaborators, Kim, Lee and coworkers have also synthesized a series of novel two-photon chromophores by utilizing dithienothiophene (DTT) as the π-center [15]. The experimental result has shown that the DTT moiety leads to an enhancement of molecular two-photon absorptivity. We hypothesized that the dramatic improvement of σ_2 values is due to this rigid, planar and polarizable fused-terthiophene structure, which provides a significant reduction of the band gap and extension of π-electron delocalization. The chemical structures and the measured σ_2 values of these chromophores are depicted in Fig. 5.

It should be noted that all the measured molecular two-photon absorption cross-section values may vary significantly if the experimental conditions such as the used organic solvent, intensity level, and pulse duration of the laser probe beam or the method of measurement are different.

4
Characterization

In this section, we describe a number of experimental techniques that can be used to probe two-photon resonances.

4.1
Non-Linear Transmission

This method involves the measurement and study of the relation between optical input and output intensities. This relationship will be linear if there is no non-linear absorption process in the medium. Therefore, a deviation from linearity indicates non-linear absorption. The theoretical modeling presents a quantitative dependence of the output intensity on the input intensity, allowing one to know whether we are dealing with two-photon or even higher-order absorption. The corresponding non-linear absorption coefficient can be determined directly by using this method [21]. The only shortcoming of this technique is that we cannot distinguish whether the multi-photon absorption under study is a sequential, step-wise absorption through real states or a direct absorption through virtual states. The experimental set-up for nanosecond non-linear absorption measurements is shown in Fig. 6. Basically, we use a linearly polarized ~800 nm pulsed laser beam as the testing beam, which is provided by a dye laser system pumped with a frequency-doubled and Q-switched Nd:YAG laser source. This laser beam is focused

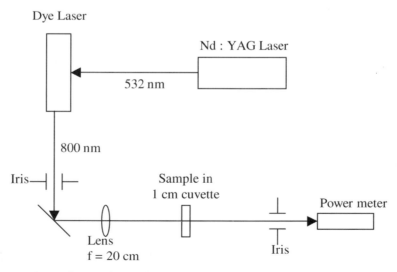

Fig. 6. Experimental set-up for nonlinear absorption measurement by nanosecond pulses

and passed through a quartz cuvette filled with the sample solution and the transmitted laser beam from the sample cell is detected by an optical power meter. Thus, if the tested sample does not have a linear absorption at 800 nm, only the transmissivity change due to pure non-linear absorption could be measured. According to the basic consideration of a TPA process, the beam intensity change along the propagation direction (z-axis) in the sample can be described as [23]:

$$dI/dz + \alpha I + \beta I^2 = 0 \tag{13}$$

where α is the attenuation coefficient due to linear absorption and scattering; β is the non-linear absorption coefficient due to TPA. The solution of Eq. (13) is

$$I(z) = \frac{I(0)e^{-\alpha z}}{1 + (\beta/\alpha)I(0) - (\beta/\alpha)I(0)e^{-\alpha z}} \tag{14}$$

where $I(0)$ is the initial intensity. In the case of small linear absorption, i.e., $\alpha z \ll 1$, Eq. (14) becomes

$$I(z) = \frac{I(0)e^{-\alpha z}}{1 + \beta z I(0)} \tag{15}$$

and the transmissivity of the non-linear medium can be written as

$$T(z) = I(z)/I(0) = \frac{e^{-\alpha z}}{1 + \beta z I(0)} = \frac{T_0}{1 + \beta z I(0)} = T_0 T_i. \tag{16}$$

Here, T_0 is the linear transmissivity independent of $I(0)$, and T_i the non-linear transmissivity dependent on $I(0)$. From Eq. (16) the non-linear absorption coefficient β can be experimentally determined easily by measuring the non-linear transmissivity T_i for a given input intensity I_0 and a given sample thickness L. If the beam is focused nearby the sample and a Gaussian transverse distribution in the non-linear medium can be assumed, the non-linear transmissivity T_i should be modified as [23]:

$$T_i = [\ln(1 + I_0 L \beta)]/I_0 L \beta \tag{17}$$

4.2
Up-Converted Fluorescence Emission

The basic idea of this method is to investigate the dependence of the fluorescence intensity on the excitation intensity because this relation determines the order of the non-linearity [8, 38–40]. For example, a quadratic dependence corresponds to a two-photon absorption process and a cubic dependence refers to a three-photon absorption. By means of a tunable excitation light source, one can also map the dispersion of the multi-photon transition, the resulting multi-photon excitation

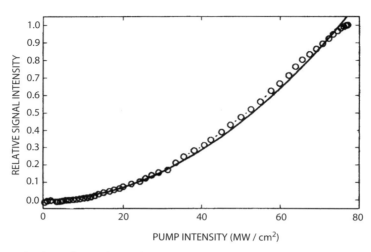

Fig. 7. Intensity dependence of two-photon induced fluorescence on the pumping light intensity. From [38]

spectra of the studied sample provide important information on the transition probabilities. One example of this method is provided by an organic crystal, DEANST ([4-N,N-diethylamino]-β-nitrostyrene) [41], as the studied sample [38]. Fig. 7 shows the relative two-photon induced fluorescence intensity as a function of the 1.06 μm pump intensity. The measured data (hollow circles) are in good agreement with the best fitting curve (the square law, $y = ax^2$), which indicates that a two-photon absorption process occurs within the sample during excitation.

4.3
Transient Absorption

This technique utilizes a pulse pump-probe experiment and monitors the absorption of a weak probe beam in the presence of a strong pump beam. Fig. 8 depicts the experimental set-up for a two-beam pump-probe experiment, which includes homodyne and heterodyne Kerr gate measurements and polarization-controlled transient absorption measurement. Generally, the input beam is produced from an amplified pulse laser system with 1 KHz repetition rate, which can produce picosecond or femtosecond pulses. This pumping light beam is divided into two beams by a beam-splitter with an intensity ratio of 30:1; therefore, the one with the stronger intensity will act as the pump and the weaker one will be the probe. The position of the sample is where these two beams focus and overlap spatially. The time delay between the pulses from these two beams is controlled by a retroreflec-

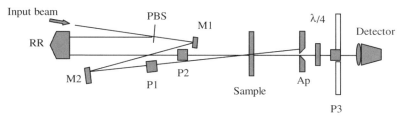

Fig. 8. Femtosecond two-beam pump-probe experimental set-up for Kerr gate and transient absorption measurement. P: polarizers; M: mirrors; PBS: Pellicle beam splitter; Ap: aperture; RR: retroreflector

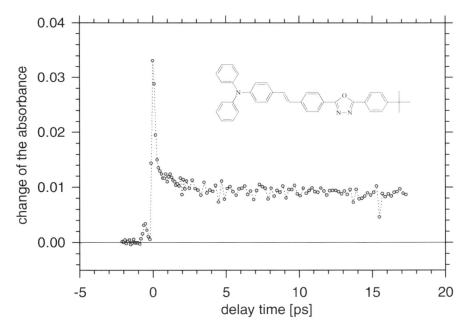

Fig. 9. Chemical structure and time-resolved two-photon induced transient absorption of 0.01 M **PRL-101** solution in chloroform

tor prism placed on a translator stage and moved with a stepper motor. When the pump beam and the probe beam temporally overlap at the sample position (focus point), there will be a sharply increased absorption of the latter because of the direct two-photon absorption. The probe pulses are gradually delayed in time with respect to the pump pulses and the transmission (or absorption) of the probe beam through the sample is measured as a function of the delay time. Absorption of the probe beam for a non-zero time delay points to absorption from excited states. Therefore, this time-resolved measurement provides information on the relative roles of direct two-photon absorption and excited states absorption [42]. Fig. 9 depicts the transient absorption signal for our newly synthesized two-photon chromophore, PRL-101, in chloroform solution by utilizing femtosecond pulses. The imaginary part of $\chi^{(3)}$ is here responsible for the pump/probe beam coupling and the fast component of the transient absorption signal (the peak). The change of the imaginary part of the linear susceptibility gives the long tail. Since there is no linear absorption at our experimental wavelength (~800 nm) of the ground state molecules, this long tail effect is due to the linear absorption of the excited molecules.

4.4
Four-Wave Mixing

In a four-wave mixing experiment, two pulses cross the medium at an angle to form interference and produce intensity modulation. This intensity modulation will lead to complex refractive index modulation, which can be probed by the diffraction of a probe beam in the time-resolved study. Therefore, one can monitor the delay and the build-up of the refractive index grating. For a material with no non-linear absorption, the detected signal will be proportional to the cubic power of the input intensities. If a studied material exhibits two-photon absorption, the detected signal will have one contribution from the coherent four-wave mixing process which has a cubic power dependence on the intensity and the other contribution from the population grating process which shows a fifth power dependence on intensity if the excited state produced by this non-linear absorption has a sufficiently long lifetime. Thus, from the dependence of the detected signal on the input intensity, one can obtain useful information to determine both the real and imaginary parts of $\chi^{(3)}$ [43].

4.5
Z-Scan Technique

The Z-scan technique, developed by Sheik-Bahae et al. [44, 45], offers a useful way to probe both the non-linear refractive index and the non-linear absorption coefficient β of a sample. In the Z-scan experiment, a laser beam is focused to a minimum waist at the focal point z_0 along the propagation direction (z-axis) of this beam. By moving the sample along the z-axis, the light intensity in the sample is varied. When the sample is placed away from the focal point, no non-linear processes can be observed in the sample because the light intensity is too low. The light intensity within the sample is increased when the sample is moved toward the focus, producing non-linear processes. The consecutive recording of the relative power transmitted through the sample as a function of the sample position pro-

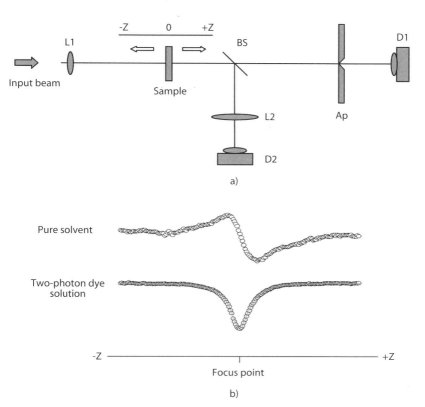

Fig. 10. (*a*) Experimental set-up for femtosecond Z-scan. BS: beam-splitter; L1,L2: lenses; D1,D2: photo-detectors. (*b*) Typical shapes of signals of pure solvent and non-linear dye solution

vides important information about the real part (the refractive index change) and the imaginary part (the intensity-dependent absorption coefficient) of $\chi^{(3)}$. Fig. 10a is the experimental set-up for the femtosecond Z-scan arrangement utilized in our laboratory. A typical signal for a pure solvent where only the refractive effect (focusing and defocusing effect) can be observed by means of the D1 detector through an aperture is shown at the top of Fig. 10b. An induced two-photon absorption of a dye solution which is measured by the D2 detector with an open aperture, a characteristic V-shape signal representing a strong non-linear absorption, is illustrated at the bottom of Fig. 10b. Here, we present one of our examples that uses the Z-scan method with femtosecond pulses to obtain effective molecular two-photon absorption cross-section (Fig. 11). It should be noticed that the σ_2 values obtained from this method are much smaller than the values obtained from nanosecond experiments. The possible reason is the excited state absorption being

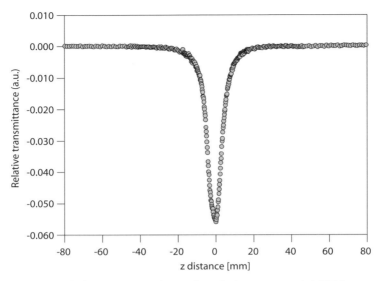

Fig. 11. V-shape signal of a 0.01 M two-photon dye solution, PRL-501, in THF from a non-linear transmission Z-scan experiment

manifested in the nanosecond pulses experiment, and thus enhancing the observed non-linear absorption [46].

5
Applications

In this section, we include several important representative applications based on two-photon absorption and up-converted emission processes either in photonics or biophotonics fields, which have been demonstrated by our research group.

5.1
Optical Power Limiting

Optical power limiting effects and devices are drawing more and more interest in the field of non-linear optics and opto-electronics. The principle of optical limiting is based on the fact that a large change in input signal will only lead to a small output change [7]. There are several different mechanisms which can lead to an optical limiting behavior. Some examples are reverse saturable absorption (RSA), two-photon absorption (TPA), non-linear refraction, and optically triggered scattering. Here, we present the optical limiting behavior achieved using a two-photon chromophore, compound **6** in Fig. 2 [9]. This compound shows great solubility in many common organic solvents and very strong two-photon absorption in the so-

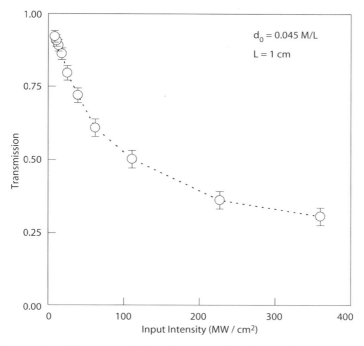

Fig. 12. Measured non-linear transmission of a 1-cm-path solution of compound **6** (of Fig. 2) in benzene of $d_0 = 0.045$ mol L^{-1} as a function of the input intensity of the ~800 nm laser beam. From [9]

lution phase. Fig. 12 shows the experimental curve for non-linear transmissivity, T_i, as a function of the input ~800 nm beam intensity in a 1-cm-path solution. One can see that the intensity-dependent, non-linear transmission of the sample dropped from ~0.93 to ~0.3 as the input intensity increased from ~10 MW/cm^2 to ~360 MW/cm^2. Based on the same experimental data, the optical power limiting behavior of the same solution sample can be presented in another way as shown in Fig. 13. The most interesting feature of Fig. 13 is that the output/input curve levels off when the input intensity increases from ~100 MW/cm^2 to ~400 MW/cm^2. This type of output/input characteristic curve can be used for optical peak power (or peak intensity) limiting and stabilization. This means that a larger fluctuation in the input peak power (or peak intensity) will lead to a much smaller output fluctuation by utilizing a non-linear absorptive medium with a sufficiently large σ_2 value. Our experimental results for optical stabilization are shown in Fig. 14. The upper curves in Fig. 14a and 14b are the instantaneous changes in peak intensity of the input and output of the ~800 nm laser pulses, while the lower curves represent the base line fluctuation of the detection system itself without laser irradia-

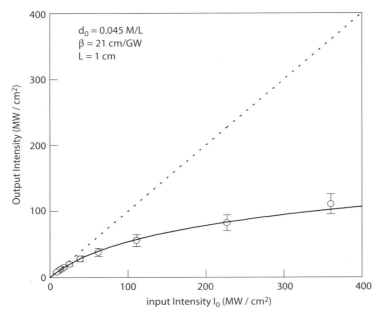

Fig. 13. Measured output intensity as a function of the input intensity based on a 1-cm-path sample solution of compound **6** in benzene with $d_0 = 0.045$ mol L^{-1}. Here, the measured transmitted intensity data are represented by hollow circles, the solid line is the theoretical curve predicted by Eq. (17), and the dotted straight line shows the behavior for a medium without non-linear absorption. From [9]

tion. Before the sample solution was used, a relative fluctuation of ±50%, as shown in Fig. 14a, was found. After passing through the solution of compound **6**, much less fluctuation for the output laser beam was observed as shown in Fig. 14b. The measured two-fold reduction of the peak intensity fluctuation is basically in agreement with the prediction from the solid curve in Fig. 13.

5.2
Up-Converted Lasing

Frequency up-converted lasing is an important research area and has become more promising in recent years because this technique provides several main advantages when compared to other coherent frequency up-conversion techniques:
1. elimination of phase-matching requirement,
2. accessibility of using the semiconductor lasers as pump sources, and
3. capability of adopting waveguide and fiber configurations.

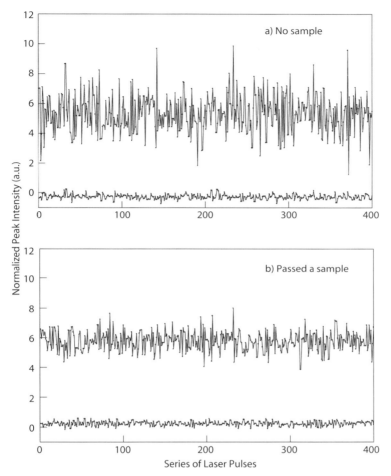

Fig. 14. (*a*) Measured instantaneous peak intensity fluctuation of the input laser pulses, and (*b*) measured instantaneous peak intensity fluctuation of the output laser pulses. The repetition rate of the laser pulses was 2 Hz, and the average input intensity level was ~250 MW cm^{-2}. From [9]

So far, there are two major technical approaches that have been used to achieve frequency up-conversion lasing. One is based on direct two-photon (or multi-photon) excitation of a gain medium, the other on sequential stepwise multi-photon excitation. The earliest report of two-photon pumped (TPP) lasing was from in a PbTe crystal at 15 K [47]. The pump wavelength was 10.6 μm, and the lasing wavelength was ~6.5 μm. Although TPP lasing action has also been observed in a number of other semiconductor crystals [48–51], a low operating temperature (10~260 K) was required. On the other hand, there are some reports of TPP lasing

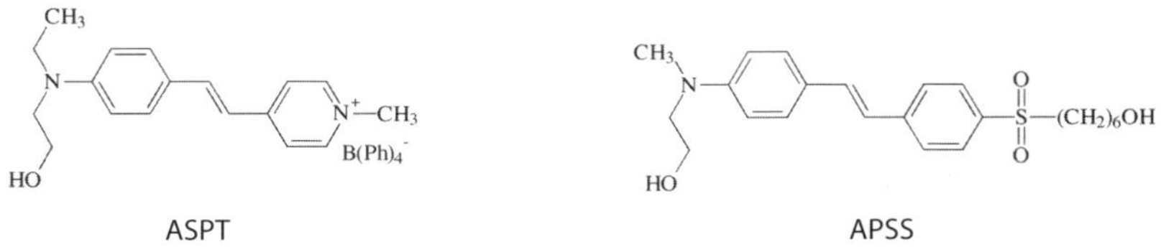

Fig. 15. Chemical structures of two-photon chromophores, ASPT and APSS, for up-converted lasing application

behavior in organic dyes with "cavity-less" lasing or superradiation (directionally amplified spontaneous emission) [52–58]. Moreover, due to the small two-photon absorptivity of commercial laser dyes, the efficiencies of TPP lasing performance reported so far have been relatively low. Two of the multifunctional chromophores synthesized in our laboratory; trans-4-[p-(N-ethyl-N-hydroxyethyl-amino)styr-yl]-N-methylpyridium tetraphenylborate (ASPT), and 4-[N-(2-hydroxyethyl)-N-(methyl)aminophenyl]-4'-(6-hydroxyhexylsulfonyl)stilbene (APSS), have been used as very effective two-photon pumped lasing media in different geometries and in different host matrices. The chemical structures of these chromophores are shown in Fig. 15. ASPT shows a strong two-photon absorption at 1.06 μm, with up-converted yellow-red fluorescence and lasing at ~600 nm [18] while APSS can be pumped at 800 nm and exhibits two-photon pumped lasing at 555 nm [16]. We have reported the first solid-state two-photon pumped "cavity" lasing in an ASPT dye-doped polymer rod [19]. Both of these chromophores have high solubility in many common organic solvents. It is also possible to dope them into the sol-gel/polymer composites and obtain two-photon pumped solid-state lasing. Furthermore, we also have achieved two-photon pumped lasing from an APSS-solution-filled hollow-fiber and the lasing threshold was as low as 0.1 mJ [20].

5.3
3-D Data Storage

The development of multimedia and electronic communication networks has triggered the need for data storage which is expected to exceed 10^{20} bits per storage media. Due to this astronomical increase in the requirement of data storage, intense research activity is going on to find alternate methods and storage media for such a large amount of data. Recently, the focus has shifted from two-dimensional to three-dimensional storage. Several approaches to three-dimensional (3-D) optical data storage, such as, holographic recording with photorefractive media and photopolymers [59–61], hole burning [62], and photon echo [63], are currently being investigated. The use of two-photon processes for optical data storage was first introduced by Rentzepis [3] and, since then, there have been several reports

also proposing the use of two-photon processes for optical data storage [24–28]. The advantages of two-photon based memory systems are:
1. volume storage with high data storage densities up to the order of 10^{12} bits/cm^3,
2. fast read/write speed,
3. feasibility of random access, and
4. low cost storage media.

Since the two-photon excitation has a quadratic dependence on the pump intensity, the excitation and subsequent photoreaction related to the writing process occurs only in the near vicinity of the focal point. An excellent axial resolution during the writing process is possible due to this property of two-photon induced processes. The basic components of a two-photon memory are:
1. a medium which can exhibit a change in its optical properties (absorbance, fluorescence, refractive index, etc.) after two-photon absorption,
2. appropriate read and write beams, and
3. a mechanism to precisely access any volume element in the medium.

We have already demonstrated 3-D data storage by using a two-photon confocal microscope (to write and read) [26]. Here we present a data storage system based on the two-photon induced spectral shift of a dye which permits the use of a simple, low-power and inexpensive CW laser for read back, thus avoiding the usage of complicated and expensive pulsed laser system [26]. To demonstrate this tech-

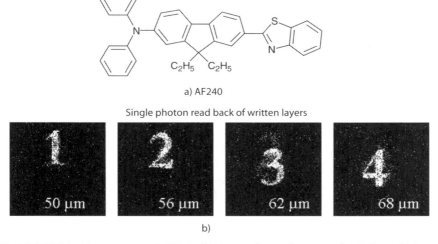

a) AF240

Single photon read back of written layers

b)

Fig. 16. (*a*) Molecular structure of AF240; (*b*) Image of one-photon readback data which was written inside of a polymeric storage medium by the two-photon process. From [26]

nique we used poly(methyl methacrylate) (PMMA) doped with AF240, a two-photon dye from U.S. Air Force, as the storage medium [26]. A tightly focused pulsed IR beam (a Ti:Sapphire laser operating at 800 nm with a pulse width 80 fs and a repetition rate of 90 MHz as light source and a high NA objective for focusing) were used to write barcodes in the polymer medium. Necessary software and hardware were developed to convert computer-generated images or barcodes into the medium. In this storage medium, the linear absorption and fluorescence properties of the written spots are red-shifted compared to the unwritten region. Here the written spots show linear absorption at 500 nm and emission at 570 nm, making it possible to use a readback system comprised of a green laser as the excitation source and confocal detection of the emission at 570 nm. Similar techniques were used to write barcodes into different polymer/dye composites utilizing the change in emission or refractive index of a two-photon dye. Based on these techniques, we were able to write multiple layers of information in a single polymer block at a vertical separation of 6 microns, and up to a depth of couple of hundred microns. Fig. 16 shows the molecular structure of AF240 and one example of the readback images of multilayer-stored information in this study.

5.4
3-D Micro-Fabrication

In recent years, three-dimensional micro-fabrication by direct laser writing or laser rapid-prototyping has been attracting a great deal of interest for its various applications to micro-electromechanical system (MEMS) [64], 3-D optical waveguide circuitry [65–67], 3-D optical data storage [3, 24–28], etc. Up to now, most of the optical circuitry sensors and actuators were fabricated by conventional lithography and an etching process on silicon substrates, which includes many complicated and time-consuming steps. On the other hand, laser rapid-prototyping enables computer-aided, single-step generation of three-dimensional structures by laser-induced photo-polymerization or photo-cross-linking. This process not only avoids the usage of photoresists or masks but also favors the complex pattern fabrication. Furthermore, because of limited depth access in bulk materials based on one-photon (or linear) absorption-induced photo-polymerization, it is necessary to perform a layer by layer fabrication.

Recently, we have developed a new approach using non-linear absorption in an organic chromophore to enhance the resolution and to access the volume of a pre-fabricated bulk material [67]. This new approach is based on the specific property of a two-photon process, which can localize the reaction or polymerization and improve the resolution of the fabricated structure. Two-photon processes can avoid the problems of linear absorption and hence the laser light can penetrate deeply into the material without loss, which gives the advantage of volume access. Fig. 17 shows some examples of fabrication of single mode and multimode chan-

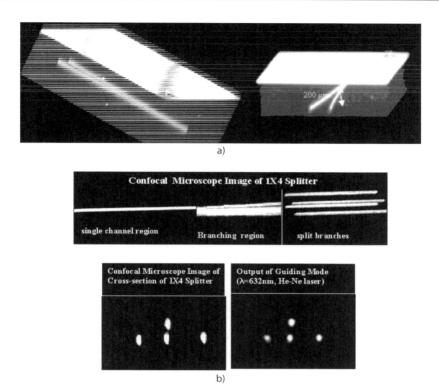

Fig. 17. Confocal microscope images of (*a*) Y-branch channel waveguide where two branches are splitting horizontally. (*b*) 1 × 4 splitter. From [68]

nel waveguides and splitters inside a pre-fabricated bulk, which were produced in our laboratory [68].

5.5
Two-Photon Fluorescence Microscopy

Confocal fluorescence microscopy is a tool commonly used to generate three-dimensional (3-D) images of biological materials such as tissues and cells [69, 70]. A laser scanning confocal microscope in fluorescent mode detects the fluorescence emission from a sample which is generally stained with a fluorescent dye (fluorophore) and excited by visible or UV laser light through single-(or one) photon excitation. The fluorophore may be chemically an integral part of the system or just physically dispersed. The emitted fluorescence, which is at a longer wavelength (towards red) than the excitation source, is detected by a photodetector, commonly a photomultiplier tube (PMT) in conjunction with a spatial filter in the form of a confocal aperture, which allows only fluorescence emitted from the focal point to pass through to the photodetector. In this way, any fluorescence that is out of focus will be eliminated. A scanning mechanism raster scans the beam over a plane of the volume to be imaged and by repeating this process over successive planes along the optical axis, a complete 3D image of the sample can be obtained. Although one-photon confocal microscopy has found wide usage, it has some serious limitations. For example, the penetration depths of UV or visible excitation light in many organic materials are small due to the extensive linear attenuation. This normally limits the depth in a material that can be probed by means of conventional confocal microscopy. Furthermore, using UV/visible light for excitation increases photobleaching and autofluorescence from either the studied samples or optical components. The advent of two-photon confocal laser scanning microscopy (2PCLSM) or two-photon laser scanning microscopy (2PLSM) has improved this situation. Two-photon fluorescence microscopy was first demonstrated by Denk et al. in 1990 [4]. This combines the unique properties of two-photon absorption with the features offered by conventional confocal microscopy. Multi-photon excitation in confocal microscopy is a non-destructive evaluation technique, which provides a significant advantage in the gain of three-dimensional resolution, image contrast, and probing depth [32–35]. Basically, in multi-photon microscopy, one uses a direct two- or three-photon pumped non-linear absorption to excite a fluorophore, which then fluoresces. In this process, the pumping wavelength is longer (towards the IR), while the emission is up-converted to the visible. Because this non-linear absorption is highly intensity-dependent, strong absorption and subsequent emission occurs very localized at the focal point (within an appropriate pumping intensity range) which provides the opportunity to increase the axial resolution, even without the spatial filtering by a confocal aperture. Since the multiphoton absorption process is relatively much weaker than the linear absorption, and the wavelength is in the near IR, the penetration depth in the sample can be very long. Therefore, one can easily study structures with depth profile and investigate the surface, the bulk. and any underlying interface. The use of infrared laser light for excitation in two-photon microscopy technology opens up the entire visible spectrum for multiple detecting channels (multi-channel microscopy). One can use different color fluorophores in different components of a system to conduct multi-color, multi-photon confocal microscopy and selectively probe different regions or sites. Here we present the most recent published data that used 2PCLSM for optical tracking of the cellular pathway in a chemotherapy which has been widely used in the treatment of cancers [31]. For this we synthesized a new dye, C625 (an APSS derivative), with strong two-photon fluorescence and functional groups for chemical attachment to many kinds of bio-molecules such as peptides. We have successfully coupled the chromophore (C625) to a chemotherapeutic agent AN-152. The chemical structure of the complex AN-152:C625 is illustrated in Fig. 18. AN-152 is made by coupling the cytotoxic agent doxorubicin

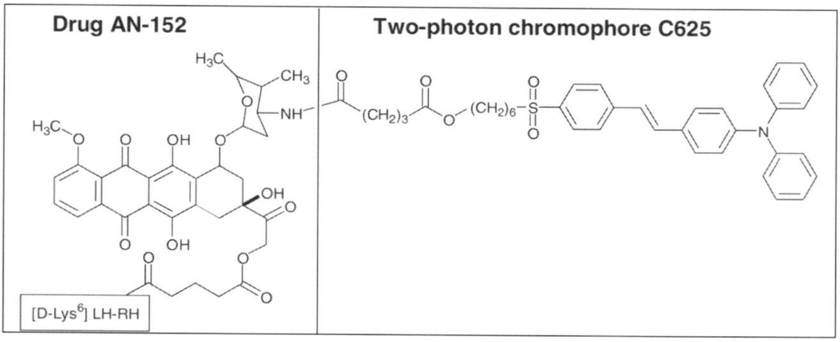

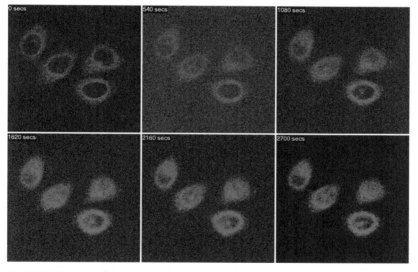

Fig. 18. Chemical structure of the complex: AN-152:C625

Fig. 19. 2PLSM image of the penetration process of the anticancer drug AN-152. Here, the breast cancer cells with up-regulated receptors are treated with AN-152:C625 complex and are continuously tracked at different times

(Dox) to the luteinizing hormone-releasing hormone (LH-RH) analogue, [D-Lys[6]] LH-RH, and it is a promising new drug used in chemotherapy [71, 72]. We used this drug-chromophore complex system to track the entry of the chemotherapeutic agent into the cancer cells by two-photon laser-scanning microscopy. Fig. 19 shows the entry pattern of the tagged drug inside a eukaryotic cancer cell. Using this technique, it is possible to identify the accumulation of the drug in different parts of the cell, with time. This approach can be further extended by coupling an efficient two-photon probe to many other biologically active molecules such as proteins, peptides, nucleic acids, etc. and track them optically either in vitro or in vivo in real-time with minimal photo-damage to the cells.

5.6
Two-Photon Photodynamic Therapy

Photodynamic therapy (PDT) is a novel treatment that involves a photosensitizer in the presence of light to produce a cytotoxic (cell death) effect on cancerous cells [73]. The modern PDT technique was pioneered by Dougherty in 1970s [74, 75]. In brief, PDT requires three elements: a photosensitizer, oxygen, and light. The photosensitizer is an efficient generator of singlet oxygen which is believed to act as a cytotoxic agent in PDT. Upon the absorption of light, the photosensitizer is excited to a short-lived singlet state from which it undergoes an intersystem crossing to a triplet state with a longer lifetime. In the presence of atmospheric oxygen, which has a triplet ground state, the most efficient reaction is the triplet-triplet annihilation. After this reaction, the photosensitizer is restored to its ground state and the singlet oxygen which is highly reactive and causes permanent damage to living tissue is generated at the same time. With the recent approval by the Food and Drug Administration (FDA) of Photofrin®, a porphyrin, as the photosensitizer, PDT has found wide application for cancer treatment. Photofrin has several absorption peaks around 400–500 nm and the strongest one is at 420 nm. However, the penetration depth of light at these wavelengths in living tissue is minimal. Thus, there is a trade-off between the penetration depth and the photosensitizer excitation efficiency. Conventionally, PDT uses the laser emission at 630–650 nm but at this wavelength, the tissue penetration is only about 2–4 mm, the photodynamic therapy effect is generally seen up to a depth 2–3 times greater. Thus, the largest attainable depth is about 15 mm, but in most cases it is less than the half of this depth. Therefore, the increase of light penetration is considered to be an important task to improve the clinical efficiency of PDT. The wavelength with better transmission (or penetration) through tissues lies around 800–1100 nm, which is in the near-infrared region and may be used for exciting the photosensitizer via two-photon absorption. Current photosensitizers themselves do not exhibit sufficiently large two-photon absorption cross-sections to provide practical significance and also the high intensities required to excite these photosensitizers by two-photon absorption may cause damage to healthy tissue. Therefore, we have proposed a new approach to PDT by using one of the efficient two-photon chromophores to be either in the proximity or chemically attached to a photosensitizer [36]. In this approach, a two-photon dye is excited by short laser pulses; it transfers the energy to a photosensitizer after being de-excited. The photosensitizer is thus excited to the singlet state from which the same sequence of energy transfer occurs as mentioned earlier and produces the singlet oxygen. A comparison of the tradi-

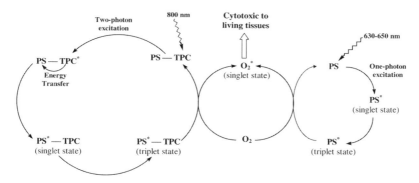

Fig. 20. Process of the conventional photodynamic therapy (right) and the newly proposed approach based on two-photon process (left). PS: photosensitizer; TPC: two-photon chromophore

tional and proposed novel approaches is depicted in Fig. 20. The preliminary in vivo studies were conducted in mice and have shown encouraging results [36]. However, further experiments are needed to investigate and increase the energy transfer efficiency.

6
Polymeric Materials

Although the two-photon chromphores mentioned above provide high two-photon absorption cross-section values and stronger frequency up-conversion emission than commercialized dyes, there are two major drawbacks for using these molecular materials for applications:
1. Photo-degradation and/or photo-bleaching caused by intense light.
2. Aggregation at high concentration.

In order to resolve this problem, several efforts have been made to develop novel polymeric materials for photonics [76–81] as they provide the following advantages:
1. Higher concentration of the absorptive and fluorescent centers without aggregation.
2. Improved optochemical and optophysical stability upon intense radiation.
3. Ease of processing and selective thermal cross-linkable to produce glassy and/or gel-type polymeric framework structures, which may be more suitable for some photonic applications.

Polymeric conjugation structures based on poly(*p*-phenyleneethynylene) (PPE) and poly(*p*-phenylenevinylene) (PPV) backbones have drawn much attention and are well-known in the light-emitting diodes research field because of their electro-luminescent properties. Through the chemical modification of their main chain structures, one can obtain various electronic structures which may lead to different spectral and optical properties.

Recently, we have reported the two-photon property and optical power limiting performance of a polymer derivative, based on the backbone of PPE [22]. PPE and its derivatives exhibit large photoluminescence quantum efficiency in both solution and solid states [82] and have been successfully used as the emission layer in electroluminescence devices [83, 84] as well as a photoluminescent polarizer in liquid crystal displays [85–87]. We have investigated the two-photon property of poly(2,5-dialkoxy-p-phenyleneethynylene) (EHO-OPPE), which is a rigid and conjugated polymer with electron-donating long chain alkyloxy functional groups on each repeat unit. Some other third-order, non-linear optical properties of this polymer have also been investigated earlier [88]. The chemical structure and optical properties of EHO-OPPE are shown in Fig. 21. The optical power limiting performance of this polymeric material is demonstrated in Fig. 22. Our group, in collaboration with Professor Jin's group in Korea, has investigated the two-photon properties of PPV-based conjugated polymers synthesized in Professor Jin's laboratory [89, 90]. In this PPV-based polymer system, the pendent groups play an important role to affect the chromophoric interaction on polymer chains and cause differences in the effective π-conjugation length as well as in the two-photon property. Fig. 23 depicts the chemical structures and our preliminary measured values of two-photon absorption cross-section of some representative polymers (including MEH-PPV, a well-known electroluminescent material as a comparison), utilizing nanosecond pulses.

For frequency up-conversion lasing application, we have synthesized a partially cross-linked, fluorescent polymer system and demonstrated the two-photon

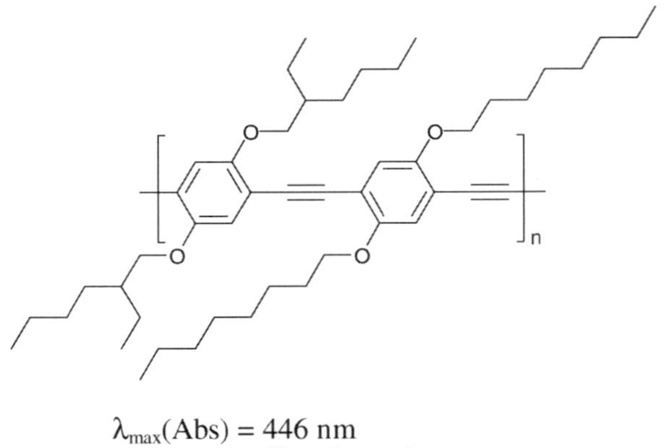

$$\lambda_{max}(Abs) = 446 \text{ nm}$$
$$\sigma_2 = 80 \times 10^{-20} \text{ cm}^4 \text{ GW}^{-1}$$

Fig. 21. Molecular structure and measured optical properties of EHO-OPPE. From [22]

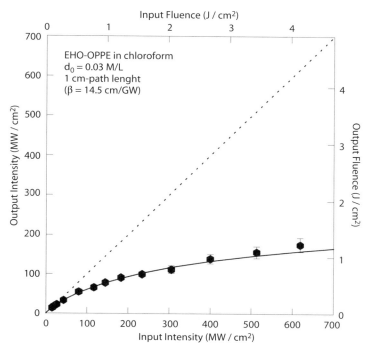

Fig. 22. Measured optical power limiting performance of a EHO-OPPE polymer solution. From [22]

pumped frequency up-converted cavity lasing phenomenon by utilizing 1064 nm as the pumping wavelength [91]. The major components involved in this homogeneous, cross-linked co-polymer are the chromophore, 4-[bis(2-hydroxyethyl)amino]-N-methylstilbazolium iodide (DHASI), the coupling agent, tolylene-2,4-diisocyanate (TDI), and the monomer, 2-hydroxyethyl methacrylate (HEMA). Compared to low-molecular weight organic dyes, this cross-linked polymeric structure provides the advantage that one can control and build up the concentration of the fluorescent centers without any aggregation problem. Furthermore, the thermal and mechanical properties are also judiciously tunable in this co-polymer system.

The chromophore, DHASI, was synthesized by coupling 4-[N,N-bis(2-hydroxyethylamino]benzaldehyde and 4-methyl-N-methylpyridium iodide under catalytic reaction conditions. To prepare this homogeneous polymer, a certain amount of DHASI was dissolved in a mixture of HEMA and DMF. The coupling reagent, TDI, was added to this mixed solution dropwise at room temperature followed by adding an extra amount of HEMA and the radical-forming agent, 1,1'-azobis(cyclohexanecarbodinitrile). This reaction mixture was then heated up to 60 °C for

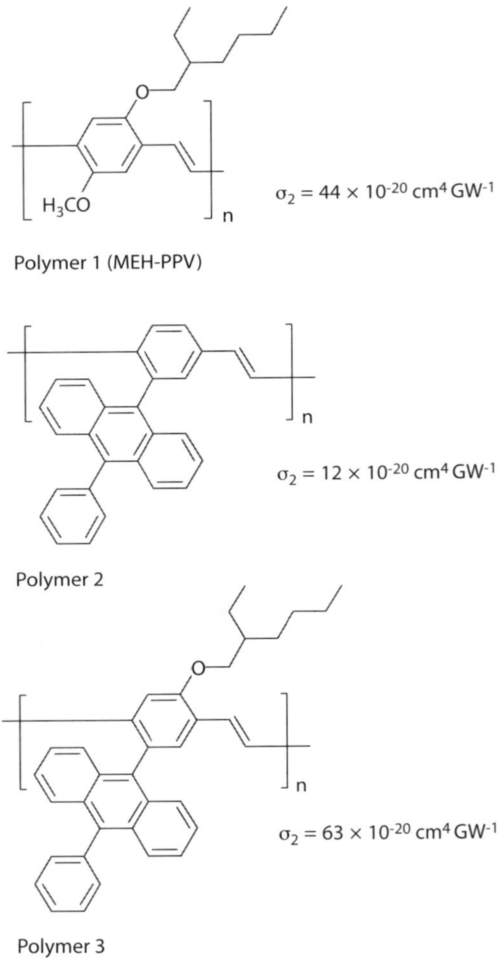

$\sigma_2 = 44 \times 10^{-20}$ cm^4 GW^{-1}

Polymer 1 (MEH-PPV)

$\sigma_2 = 12 \times 10^{-20}$ cm^4 GW^{-1}

Polymer 2

$\sigma_2 = 63 \times 10^{-20}$ cm^4 GW^{-1}

Polymer 3

Fig. 23. Chemical structures and measured two-photon absorptivities of polymer samples based on a PPV backbone

148 hours for polymerization. The molecular structure of DHASI and the repeat unit of this cross-linked polymer, poly(DHASI-HEMA), are shown in Figs. 24a and b, respectively.

In order to test the two-photon pumped, up-conversion lasing performance of this polymer, a polymer rod of 13 mm length with two polished ends was prepared. This polymer rod was placed in a 20 mm long glass cell filled with a refractive index-matching transparent liquid to reduce the influence of the non-optical polishing quality of the two ends [92]. The 1064 nm pumping beam provided by

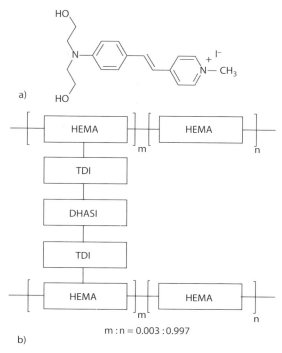

a)

b)

m : n = 0.003 : 0.997

Fig. 24. Chemical structure of the chromophore DHASI (*a*) and a schematic of the repeat unit of the poly(DHASI-HEMA) system (*b*). From [91]

a Q-switched pulsed Nd:YAG laser source was focused by a lens at the point slightly before the polymer rod to avoid any possible damage to the studied sample. The spectral structure of the lasing output was measured by using a grating spectrograph in conjunction with an optical multi-channel analyzer system. The spectra of the single-photon-induced emission and the frequency up-conversion cavity lasing at various pumping energy levels of this polymer rod are shown in Fig. 25.

7
Summary

Novel organic structures either in molecular or polymeric forms with highly efficient two-photon absorbing and up-converting properties have currently opened up various potential applications in photonics and biophotonics including optical power limiting, up-converted lasing, 3-D data storage, 3-D microfabrication, two-photon fluorescence microscopy, bioimaging, and two-photon photodynamic therapy as described in detail in this article. The development of more advanced equipment and characterization techniques for the study of optical non-linearities

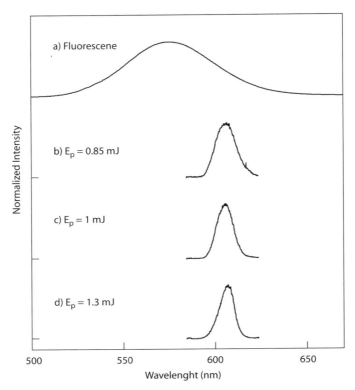

Fig. 25. Single-photon-induced emission spectrum (*curve a*) and two-photon pumped lasing spectra at various energy levels (*curves b–d*) of the poly(DHASI-HEMA) system. From [91]

will not only help synthetic chemists gain insight into structure-property relationships and subsequently contribute to the strategy of molecular design but will also inspire materials scientists to explore new applications based on the two-photon absorption phenomenon.

Acknowledgement. The research reported here was supported by the Chemistry and Life Sciences of the Air Force Office of Scientific Research and the Polymer Branch of the Air Force Research Laboratory in Dayton, Ohio.

References

1. Göppert-Mayer M (1931) Ann Phys 9:273
2. Kaiser W, Garret CGB (1961) Phys Rev Lett 7:229
3. Parthenopoulos DA, Rentzepis PM (1989) Science 245:843
4. Denk W, Strickler JH, Webb WW (1990) Science 248:73
5. Prasad PN, Williams DJ (1991) Introduction to Nonlinear Optical Effects in Molecules and Polymers. Wiley, New York

6. Bhawalkar JD, He GS, Prasad PN (1996) Rep Prog Phys 59:1041
7. He GS, Gvishi R, Prasad PN, Reinhardt BA (1995) Opt Commun 117:133
8. He GS, Bhawalkar JD, Zhao CF, Prasad PN (1995) Appl Phys Lett 67:2433
9. He GS, Yuan L, Cheng N, Bhawalkar JD, Prasad PN, Brott LL, Clarson SJ, Reinhardt BA (1997) J Opt Soc Am B 14:1079
10. Reinhardt BA, Brott LL, Clarson SJ, Dillard AG, Bhatt JC, Kannan R, Yuan L, He GS, Prasad PN (1998) Chem Mater 10:1863
11. Chung S-J, Kim K-S, Lin T-C, He GS, Swiatkiewicz J, Prasad PN (1999) J Phys Chem B 103:10741
12. Andronov A, Fréchet JN, He GS, Kim K-S, Chung S-J, Swiatkiewicz J, Prasad PN (2000) Chem Mater 12:2838
13. Ehrlich JB, Wu X-L, Lee I-YS, Hu Z-Y, Rockel H, Marder SR, Perry JW (1997) Opt Lett 22:1843
14. Albota M, Beljonne D, Bredas J-L, Ehrlich JE, Fu J-Y, Heikal AA, Hess SE, Kogej T, Levin MD, Marder SR, McCord-Maughon D, Perry JW, Rockel H, Rumi M, Subramaniam G, Webb WW, Wu X-L, Xu C (1998) Science 281:1653
15. Kim OK, Lee KS, Woo HY, Kim KS, He GS, Swiatkiewicz J, Prasad PN (2000) Chem Mater 12:284
16. Bhawalkar JD, He GS, Park CK, Zhao CF, Ruland G, Prasad PN (1996) Opt Commun 124:33
17. Zhao CF, Gvishi R, Narang U, Ruland G, Prasad PN (1996) J Phys Chem 100:4526
18. He GS, Bhawalkar JD, Zhao CF, Prasad PN (1996) IEEE J. Quantum Electron 32:749
19. He GS, Zhao CF, Bhawalkar JD, Prasad PN (1995) Appl Phys Lett 67:3703
20. He GS, Bhawalkar JD, Zhao CF, Park C-K, Prasad PN (1995) Opt Lett 20:2393
21. He GS, Xu GC, Prasad PN, Reinhardt BA, Bhatt JC, McKellar R, Dillard AG (1995) Opt Lett 20:435
22. He GS, Weder C, Smith P, Prasad PN (1998) J Quantum Electron. 34:2279
23. Tutt LW, Boggess TF (1993) Prog. Quantum Electron 17:299
24. Dvormikov AS, Rentzepis PM (1997) Opt Commun 136:1
25. Strickler JH, Webb WW (1991) Opt Commun 16:1780
26. Pudavar HE, Joshi MP, Prasad PN, Reianhardt BA (1999) Appl Phys Lett 74:1338
27. Parthenopoulos DA, Rentzepis PM (1990) J Appl Phys 68:5814
28. Dvormikov AS, Rentzepis PM (1995) Opt Commun 119:341
29. Gura T (1997) Science 276:1988
30. Bhawalkar JD, Swiatkiewicz J, Prasad PN, Pan SJ, Shin A, Samarabandu JK, Cheng PC, Reinhardt BA (1997) Polymer 38:4551
31. Wang X, Krebs LJ, Al-Nuri M, Pudavar HE, Ghosal S, Liebow C, Nagy AA, Schally AV, Prasad PN (1999) Proc Natl Acad Sci USA 96:11081
32. Bhawalkar JD, Swiatkiewicz J, Pan SJ, Samarabandu JK, Liou WS, He GS, Berezney R, Cheng PC, Prasad PN (1996) Scanning 18:562
33. Bhawalkar JD, Swiatkiewicz J, He GS, Prasad PN, Pan SJ, Samarabandu JK, Liou WS, Cheng PC (1996) CLEO'96 Technical Digest Series 9:19
34. Bhawalkar JD, Shih A, Pan SJ, Liou WS, Swiatkiewicz J, Reinhardt BA, Prasad PN, Cheng PC (1996) J Bioimaging 4:168
35. Cheng PC, Pan SJ, Bhawalkar JD, Swiatkiewicz J, Samarabandu JK, Liou WS, He GS, Prasad PN (1996) Scanning 18:148
36. Bhawalkar JD, Kumar ND, Zhao CF, Prasad PN (1997) J Clinical Laser Medicine & Surgery 37:510
37. Dick B, Hochstrasser RM, Trommmsdorf HP (1987) In: Chemla DS, Zyss J (eds), Resonant Molecular Optics: Nonlinear Optical Properties of Organic Molecules and Crystals, vol 12. Academic, New York, p 159
38. He GS, Zieba J, Bradshaw JT, Kazmierczak MR, Prasad PN (1993) Opt Commun 104:102
39. Bhawalkar JD, He GS, Prasad PN (1995) Opt Commun 199:587
40. He GS, Bhawalkar JD, Prasad PN (1995) Opt Lett 20:1524

41. Kurihara T, Kanbara H, Kobayashi H, Kubodera K, Matsumoto S, Kaino T (1991) Opt Commun 84:149
42. Pang Y, Samoc M, Prasad PN (1991) J Chem Phys 94:5282
43. Zhao MT, Cui Y, Samoc M, Prasad PN, Unroe MR, Reinhardt BA (1991) J Chem Phys 3:864
44. Sheik-Bahae M, Said AA, Van Stryland EW (1989) Opt Lett 14:955
45. Shiek-Bahae M, Said AA, Wei TH, Hagan DJ, Van Stryland EW (1990) IEEE J Quantum Electron 26:760
46. Swiatkiewicz J, Prasad PN, Reinhardt BA (1998) Opt Commun 157:135
47. Patelm CKN, Fleur PA, Slusher RE, Frisch HL (1966) Phys Rev Lett 16:971
48. Yoshida T, Miyasaki K, Fujisawa K (1975) Jpn J Appl Phys 14:1987
49. Gribkovskii VP (1979) Sov J Quantum Electron 9:1305
50. Gao WL, Vaucher AM, Ling JD, Lee CH (1982) Proc SPIE Int Soc Opt Eng 322:27
51. Yang XH (1993) Appl Phys Lett 62:1071
52. Rapp W, Gronau B (1971) Chem Phys Lett 8:529
53. Topp MR, Rentzepis PM (1971) Phys Rev A 3:358
54. Rubinov AN, Richardson MC, Sala K JAA (1975) Appl Phys Lett 27:358
55. Prokhorenko VI, Tikhonov EA, Shpak MT (1981) Sov J Quantum Electron 11:139
56. Qiu P, Penzkofer A (1989) Appl Phys B 48:115
57. Zaporozhchenko VA, Kachinskii AV, Korol'kov MV, Chenkhov OV (1989) Sov J Quantum Electron 19:1179
58. Kwok AS (1992) Opt Lett 17:1435
59. Gunter P, Huignard JP (1988) Photorefractive Materials and their Applications I. Springer, Berlin, Heidelberg, New York,
60. Hesselink L, Bashaw MC (1993) Opt Quantum Electron 25:611
61. Liphardt M, Goonesekera A, Jones B, Ducharme S, Takacs JM, Zhang L (1994) Science 263:367
62. Moerner WE (1987) Persistent Spectral Hole Burning: Science and Applications. Springer, Berlin, Heidelberg, New York
63. Kachru R, Kim MK (1989) Opt Lett 28:2186
64. Maruo S, Nkamura O, Kawata S (1997) Opt Lett 22:132
65. Wu ES, Strickler JH, Harrell WR, Webb WW (1992) SPIE Proc 1674:776
66. Frisken SJ (1993) Opt Lett 18:1035
67. Joshi MP, Pudavar HE, Swiatkiewicz J, Prasad PN, Reianhardt BA (1999) Appl Phys Lett 74:170
68. Min Y-H, Pudavar HE, Prasad PN, Bauer J, Vaia R (unpublished results)
69. White JG, Amos WB, Fordham M (1987) J Cell Biology 105:41
70. Wilson T, Sheppard C (1984) Theory and Practice of Scanning Optical Microscopy. Academic Press, London
71. Nagy A, Schally AV, Armatis P, Szepeshazi K, Halmos G, Kovacs M, Zarandi M, Groot K, Miyazaki M, Jungwirth A, Horvath J (1996) Proc Natl Acad Sci USA 93:7269
72. Schally AV, Nagy A (1999) Eur J Endocrinol 141:1
73. Fisher AMR, Murphree AL, Gomer CJ (1995) Lasers Surg Med 17:2
74. Dougherty TJ, Grindley G, Flel R (1974) J Nat. Cancer Inst 55:115
75. Dougherty TJ, Kaufman JE, Goldbarb A, Weishaupt KR, Boyle D, Mittleman A (1978) Cancer Res 38:2628
76. Moses D (1992) Appl Phys Lett 60:3215
77. Ferrer ML, Acuna AU, Amart-Guerri F, Costela A, Figuera JM, Florido F, Sastre R (1994) Appl Opt 33:2266
78. Brouwer H-J, Krasnikov VV, Hilberer A, Wildeman J, Hadziioannou G (1995) Appl Phys Lett 66:3404
79. Costela A, Garcia-Moreno I, Figuera JM (1996) J Appl Phys 80:3167
80. Hide F, Schwartz BJ, Diaz-Garcia MA, Heeger AJ (1996) Chem Phys Lett 256:424
81. Tessler N, Denton GJ, Friend RH (1996) Nature 382:695

82. Weder C, Wrighton MS (1996) Macromolecules 29:5157
83. Swanson LS, Lu F, Shinar J, Ding YW, Barton TJ (1993) Proc SPIE 1910:101
84. Montali A, Weder C, Smith P (1997) Proc. SPIE 3148:298
85. Weder C, Sarwa C, Bastiaansen C, Smith P (1997) Adv Mater 9:1035
86. Weder C, Sarwa C, Montali A, Bastiaansen C, Smith P (1998) Science 279:835
87. Montali A, Bastiaansen C, Smith PS, Weder C (1998) Nature 392:261
88. Weder C, Wrighton MS, Spreiter R, Bosshard C, Gunter P (1996) J Phys Chem 100:18931
89. Chung S-J, Jin J-I, Lee C-H, Lee C-E (1998) Adv Matr 10:684
90. Kim Y-H, Jeoung S-C, Kim D, Chung S-J, Jin J-I (2000) Chem Matr 12:1067
91. He GS, Kim K-S, Yuan L, Cheng N, Prasad PN (1997) Appl Phys Lett 71:1619
92. He GS, Bhawalkar JD, Zhao CF, Park CK, Prasad PN (1996) Appl Phys Lett 8:3549

Received June 2001

Author Index Volumes 101–161

Author Index Volumes 1–100 see Volume 100

de, Abajo, J. and de la Campa, J.G.: Processable Aromatic Polyimides. Vol. 140, pp. 23-60.
Adolf, D. B. see Ediger, M. D.: Vol. 116, pp. 73-110.
Aharoni, S. M. and *Edwards, S. F.*: Rigid Polymer Networks. Vol. 118, pp. 1-231.
Albertsson, A.-C., Varma, I. K.: Aliphatic Polyesters: Synthesis, Properties and Applications. Vol. 157, pp. 99–138.
Albertsson, A.-C. see Edlund, U.: Vol. 157, pp. 53-98.
Albertsson, A.-C. see Söderqvist Lindblad, M.: Vol. 157, pp. 139–161.
Albertsson, A.-C. see Stridsberg, K. M.: Vol. 157, pp. 27–51.
Améduri, B., Boutevin, B. and *Gramain, P.*: Synthesis of Block Copolymers by Radical Polymerization and Telomerization. Vol. 127, pp. 87-142.
Améduri, B. and *Boutevin, B.*: Synthesis and Properties of Fluorinated Telechelic Monodispersed Compounds. Vol. 102, pp. 133-170.
Amselem, S. see Domb, A. J.: Vol. 107, pp. 93-142.
Andrady, A. L.: Wavelenght Sensitivity in Polymer Photodegradation. Vol. 128, pp. 47-94.
Andreis, M. and *Koenig, J. L.*: Application of Nitrogen-15 NMR to Polymers. Vol. 124, pp. 191-238.
Angiolini, L. see Carlini, C.: Vol. 123, pp. 127-214.
Anseth, K. S., Newman, S. M. and *Bowman, C. N.*: Polymeric Dental Composites: Properties and Reaction Behavior of Multimethacrylate Dental Restorations. Vol. 122, pp. 177-218.
Antonietti, M. see Cölfen, H.: Vol. 150, pp. 67-187.
Armitage, B. A. see O'Brien, D. F.: Vol. 126, pp. 53-58.
Arndt, M. see Kaminski, W.: Vol. 127, pp. 143-187.
Arnold Jr., F. E. and *Arnold, F. E.*: Rigid-Rod Polymers and Molecular Composites. Vol. 117, pp. 257-296.
Arora, M. see Kumar, M.N.V.R.: Vol. 160, pp. 45-118.
Arshady, R.: Polymer Synthesis via Activated Esters: A New Dimension of Creativity in Macromolecular Chemistry. Vol. 111, pp. 1-42.

Bahar, I., Erman, B. and *Monnerie, L.*: Effect of Molecular Structure on Local Chain Dynamics: Analytical Approaches and Computational Methods. Vol. 116, pp. 145-206.
Ballauff, M. see Dingenouts, N.: Vol. 144, pp. 1-48.
Baltá-Calleja, F. J., González Arche, A., Ezquerra, T. A., Santa Cruz, C., Batallón, F., Frick, B. and *López Cabarcos, E.*: Structure and Properties of Ferroelectric Copolymers of Poly(vinylidene) Fluoride. Vol. 108, pp. 1-48.
Barnes, M. D. see Otaigbe, J.U.: Vol. 154, pp. 1-86.
Barshtein, G. R. and *Sabsai, O. Y.*: Compositions with Mineralorganic Fillers. Vol. 101, pp.1-28.
Baschnagel, J., Binder, K., Doruker, P., Gusev, A. A., Hahn, O., Kremer, K., Mattice, W. L., Müller-Plathe, F., Murat, M., Paul, W., Santos, S., Sutter, U. W., Tries, V.: Bridging the Gap Between Atomistic and Coarse-Grained Models of Polymers: Status and Perspectives. Vol. 152, pp. 41-156.
Batallán, F. see Baltá-Calleja, F. J.: Vol. 108, pp. 1-48.
Batog, A. E., Pet'ko, I. P., Penczek, P.: Aliphatic-Cycloaliphatic Epoxy Compounds and Polymers. Vol. 144, pp. 49-114.

Barton, J. see Hunkeler, D.: Vol. 112, pp. 115-134.

Bell, C. L. and *Peppas, N. A.*: Biomedical Membranes from Hydrogels and Interpolymer Complexes. Vol. 122, pp. 125-176.

Bellon-Maurel, A. see Calmon-Decriaud, A.: Vol. 135, pp. 207-226.

Bennett, D. E. see O'Brien, D. F.: Vol. 126, pp. 53-84.

Berry, G.C.: Static and Dynamic Light Scattering on Moderately Concentraded Solutions: Isotropic Solutions of Flexible and Rodlike Chains and Nematic Solutions of Rodlike Chains. Vol. 114, pp. 233-290.

Bershtein, V. A. and *Ryzhov, V. A.*: Far Infrared Spectroscopy of Polymers. Vol. 114, pp. 43-122.

Bigg, D. M.: Thermal Conductivity of Heterophase Polymer Compositions. Vol. 119, pp. 1-30.

Binder, K.: Phase Transitions in Polymer Blends and Block Copolymer Melts: Some Recent Developments. Vol. 112, pp. 115-134.

Binder, K.: Phase Transitions of Polymer Blends and Block Copolymer Melts in Thin Films. Vol. 138, pp. 1-90.

Binder, K. see Baschnagel, J.: Vol. 152, pp. 41-156.

Bird, R. B. see Curtiss, C. F.: Vol. 125, pp. 1-102.

Biswas, M. and *Mukherjee, A.*: Synthesis and Evaluation of Metal-Containing Polymers. Vol. 115, pp. 89-124.

Biswas, M. and *Sinha Ray, S.*: Recent Progress in Synthesis and Evaluation of Polymer-Montmorillonite Nanocomposites. Vol. 155, pp. 167-221.

Bolze, J. see Dingenouts, N.: Vol. 144, pp. 1-48.

Bosshard, C.: see Gubler, U.: Vol. 158, pp. 123-190.

Boutevin, B. and *Robin, J. J.*: Synthesis and Properties of Fluorinated Diols. Vol. 102. pp. 105-132.

Boutevin, B. see Amédouri, B.: Vol. 102, pp. 133-170.

Boutevin, B. see Améduri, B.: Vol. 127, pp. 87-142.

Bowman, C. N. see Anseth, K. S.: Vol. 122, pp. 177-218.

Boyd, R. H.: Prediction of Polymer Crystal Structures and Properties. Vol. 116, pp. 1-26.

Briber, R. M. see Hedrick, J. L.: Vol. 141, pp. 1-44.

Bronnikov, S. V., Vettegren, V. I. and *Frenkel, S. Y.*: Kinetics of Deformation and Relaxation in Highly Oriented Polymers. Vol. 125, pp. 103-146.

Brown, H. R. see Creton, C.: Vol. 156, pp. 53-135.

Bruza, K. J. see Kirchhoff, R. A.: Vol. 117, pp. 1-66.

Budkowski, A.: Interfacial Phenomena in Thin Polymer Films: Phase Coexistence and Segregation. Vol. 148, pp. 1-112.

Burban, J. H. see Cussler, E. L.: Vol. 110, pp. 67-80.

Burchard, W.: Solution Properties of Branched Macromolecules. Vol. 143, pp. 113-194.

Calmon-Decriaud, A. Bellon-Maurel, V., Silvestre, F.: Standard Methods for Testing the Aerobic Biodegradation of Polymeric Materials. Vol 135, pp. 207-226.

Cameron, N. R. and *Sherrington, D. C.*: High Internal Phase Emulsions (HIPEs)-Structure, Properties and Use in Polymer Preparation. Vol. 126, pp. 163-214.

de la Campa, J. G. see de Abajo, , J.: Vol. 140, pp. 23-60.

Candau, F. see Hunkeler, D.: Vol. 112, pp. 115-134.

Canelas, D. A. and *DeSimone, J. M.*: Polymerizations in Liquid and Supercritical Carbon Dioxide. Vol. 133, pp. 103-140.

Canva, M., Stegeman, G. I.: Quadratic Parametric Interactions in Organic Waveguides. Vol. 158, pp. 87-121.

Capek, I.: Kinetics of the Free-Radical Emulsion Polymerization of Vinyl Chloride. Vol. 120, pp. 135-206.

Capek, I.: Radical Polymerization of Polyoxyethylene Macromonomers in Disperse Systems. Vol. 145, pp. 1-56.

Capek, I.: Radical Polymerization of Polyoxyethylene Macromonomers in Disperse Systems. Vol. 146, pp. 1-56.

Capek, I. and *Chern, C.-S.*: Radical Polymerization in Direct Mini-Emulsion Systems. Vol. 155, pp. 101-166.

Carlesso, G. see Prokop, A.: Vol. 160, pp. 119–174.

Carlini, C. and *Angiolini, L.*: Polymers as Free Radical Photoinitiators. Vol. 123, pp. 127-214.

Carter, K. R. see Hedrick, J. L.: Vol. 141, pp. 1-44.

Casas-Vazquez, J. see Jou, D.: Vol. 120, pp. 207-266.

Chandrasekhar, V.: Polymer Solid Electrolytes: Synthesis and Structure. Vol 135, pp. 139-206.

Chang, J.Y. see Han, M. J.: Vol. 153, pp. 1-36.

Charleux, B., Faust R.: Synthesis of Branched Polymers by Cationic Polymerization. Vol. 142, pp. 1-70.

Chen, P. see Jaffe, M.: Vol. 117, pp. 297-328.

Chern, C.-S. see Capek, I.: Vol. 155, pp. 101-166.

Choe, E.-W. see Jaffe, M.: Vol. 117, pp. 297-328.

Chow, T. S.: Glassy State Relaxation and Deformation in Polymers. Vol. 103, pp. 149-190.

Chung, S.-J. see Lin, T.-C.: Vol. 161, pp. 157-193

Chung, T.-S. see Jaffe, M.: Vol. 117, pp. 297-328.

Cölfen, H. and *Antonietti, M.*: Field-Flow Fractionation Techniques for Polymer and Colloid Analysis. Vol. 150, pp. 67-187.

Comanita, B. see Roovers, J.: Vol. 142, pp. 179-228.

Connell, J. W. see Hergenrother, P. M.: Vol. 117, pp. 67-110.

Creton, C., Kramer, E. J., Brown, H. R., Hui, C.-Y.: Adhesion and Fracture of Interfaces Between Immiscible Polymers: From the Molecular to the Continuum Scale. Vol. 156, pp. 53-135.

Criado-Sancho, M. see Jou, D.: Vol. 120, pp. 207-266.

Curro, J.G. see Schweizer, K.S.: Vol. 116, pp. 319-378.

Curtiss, C. F. and *Bird, R. B.*: Statistical Mechanics of Transport Phenomena: Polymeric Liquid Mixtures. Vol. 125, pp. 1-102.

Cussler, E. L., Wang, K. L. and *Burban, J. H.*: Hydrogels as Separation Agents. Vol. 110, pp. 67-80.

Dalton, L. Nonlinear Optical Polymeric Materials: From Chromophore Design to Commercial Applications. Vol. 158, pp. 1-86.

Davidson, J.M. see Prokop, A.: Vol. 160, pp.119–174.

DeSimone, J. M. see Canelas D. A.: Vol. 133, pp. 103-140.

DiMari, S. see Prokop, A.: Vol. 136, pp. 1-52.

Dimonie, M. V. see Hunkeler, D.: Vol. 112, pp. 115-134.

Dingenouts, N., Bolze, J., Pötschke, D., Ballauf, M.: Analysis of Polymer Latexes by Small-Angle X-Ray Scattering. Vol. 144, pp. 1-48.

Dodd, L. R. and *Theodorou, D. N.*: Atomistic Monte Carlo Simulation and Continuum Mean Field Theory of the Structure and Equation of State Properties of Alkane and Polymer Melts. Vol. 116, pp. 249-282.

Doelker, E.: Cellulose Derivatives. Vol. 107, pp. 199-266.

Dolden, J. G.: Calculation of a Mesogenic Index with Emphasis Upon LC-Polyimides. Vol. 141, pp. 189 -245.

Domb, A. J., Amselem, S., Shah, J. and *Maniar, M.*: Polyanhydrides: Synthesis and Characterization. Vol.107, pp. 93-142.

Domb, A.J. see Kumar, M.N.V.R.: Vol. 160, pp. 45–118.

Doruker, P. see Baschnagel, J.: Vol. 152, pp. 41-156.

Dubois, P. see Mecerreyes, D.: Vol. 147, pp. 1-60.

Dubrovskii, S. A. see Kazanskii, K. S.: Vol. 104, pp. 97-134.

Dunkin, I. R. see Steinke, J.: Vol. 123, pp. 81-126.

Dunson, D. L. see McGrath, J. E.: Vol. 140, pp. 61-106.

Eastmond, G. C.: Poly(ε-caprolactone) Blends. Vol.149, pp. 59-223.

Economy, J. and *Goranov, K.*: Thermotropic Liquid Crystalline Polymers for High Performance Applications. Vol. 117, pp. 221-256.

Ediger, M. D. and *Adolf, D. B.*: Brownian Dynamics Simulations of Local Polymer Dynamics. Vol. 116, pp. 73-110.

Edlund, U. Albertsson, A.-C.: Degradable Polymer Microspheres for Controlled Drug Delivery. Vol. 157, pp. 53-98.

Edwards, S. F. see Aharoni, S. M.: Vol. 118, pp. 1-231.

Endo, T. see Yagci, Y.: Vol. 127, pp. 59-86.

Engelhardt, H. and *Grosche, O.*: Capillary Electrophoresis in Polymer Analysis. Vol. 150, pp. 189-217.

Erman, B. see Bahar, I.: Vol. 116, pp. 145-206.

Ewen, B, Richter, D.: Neutron Spin Echo Investigations on the Segmental Dynamics of Polymers in Melts, Networks and Solutions. Vol. 134, pp. 1-130.

Ezquerra, T. A. see Baltá-Calleja, F. J.: Vol. 108, pp. 1-48.

Faust, R. see Charleux, B: Vol. 142, pp. 1-70.

Fekete, E see Pukánszky, B: Vol. 139, pp. 109-154.

Fendler, J.H.: Membrane-Mimetic Approach to Advanced Materials. Vol. 113, pp. 1-209.

Fetters, L. J. see Xu, Z.: Vol. 120, pp. 1-50.

Förster, S. and *Schmidt, M.*: Polyelectrolytes in Solution. Vol. 120, pp. 51-134.

Freire, J.J.: Conformational Properties of Branched Polymers: Theory and Simulations. Vol. 143, pp. 35-112.

Frenkel, S. Y. see Bronnikov, S. V.: Vol. 125, pp. 103-146.

Frick, B. see Baltá-Calleja, F. J.: Vol. 108, pp. 1-48.

Fridman, M. L.: see Terent´eva, J. P.: Vol. 101, pp. 29-64.

Fukui, K. see Otaigbe, J. U.: Vol. 154, pp. 1-86.

Funke, W.: Microgels-Intramolecularly Crosslinked Macromolecules with a Globular Structure. Vol. 136, pp. 137-232.

Galina, H.: Mean-Field Kinetic Modeling of Polymerization: The Smoluchowski Coagulation Equation. Vol. 137, pp. 135-172.

Ganesh, K. see Kishore, K.: Vol. 121, pp. 81-122.

Gaw, K. O. and *Kakimoto, M.*: Polyimide-Epoxy Composites. Vol. 140, pp. 107-136.

Geckeler, K. E. see Rivas, B.: Vol. 102, pp. 171-188.

Geckeler, K. E.: Soluble Polymer Supports for Liquid-Phase Synthesis. Vol. 121, pp. 31-80.

Gehrke, S. H.: Synthesis, Equilibrium Swelling, Kinetics Permeability and Applications of Environmentally Responsive Gels. Vol. 110, pp. 81-144.

de Gennes, P.-G.: Flexible Polymers in Nanopores. Vol. 138, pp. 91-106.

Giannelis, E.P., Krishnamoorti, R., Manias, E.: Polymer-Silicate Nanocomposites: Model Systems for Confined Polymers and Polymer Brushes. Vol. 138, pp. 107-148.

Godovsky, D. Y.: Device Applications of Polymer-Nanocomposites. Vol. 153, pp. 163-205.

Godovsky, D. Y.: Electron Behavior and Magnetic Properties Polymer-Nanocomposites. Vol. 119, pp. 79-122.

González Arche, A. see Baltá-Calleja, F. J.: Vol. 108, pp. 1-48.

Goranov, K. see Economy, J.: Vol. 117, pp. 221-256.

Gramain, P. see Améduri, B.: Vol. 127, pp. 87-142.

Grest, G.S.: Normal and Shear Forces Between Polymer Brushes. Vol. 138, pp. 149-184.

Grigorescu, G, Kulicke, W.-M.: Prediction of Viscoelastic Properties and Shear Stability of Polymers in Solution. Vol. 152, p. 1-40.

Grosberg, A. and *Nechaev, S.*: Polymer Topology. Vol. 106, pp. 1-30.

Grosche, O. see Engelhardt, H.: Vol. 150, pp. 189-217.

Grubbs, R., Risse, W. and *Novac, B.*: The Development of Well-defined Catalysts for Ring-Opening Olefin Metathesis. Vol. 102, pp. 47-72.

Gubler, U., Bosshard, C.: Molecular Design for Third-Order Nonlinear Optics. Vol. 158, pp. 123-190.

van Gunsteren, W. F. see Gusev, A. A.: Vol. 116, pp. 207-248.

Gusev, A. A., Müller-Plathe, F., van Gunsteren, W. F. and *Suter, U. W.*: Dynamics of Small Molecules in Bulk Polymers. Vol. 116, pp. 207-248.

Gusev, A. A. see Baschnagel, J.: Vol. 152, pp. 41-156.
Guillot, J. see Hunkeler, D.: Vol. 112, pp. 115-134.
Guyot, A. and *Tauer, K.*: Reactive Surfactants in Emulsion Polymerization. Vol. 111, pp. 43-66.

Hadjichristidis, N., Pispas, S., Pitsikalis, M., Iatrou, H., Vlahos, C.: Asymmetric Star Polymers Synthesis and Properties. Vol. 142, pp. 71-128.
Hadjichristidis, N. see Xu, Z.: Vol. 120, pp. 1-50.
Hadjichristidis, N. see Pitsikalis, M.: Vol. 135, pp. 1-138.
Hahn, O. see Baschnagel, J.: Vol. 152, pp. 41-156.
Hakkarainen, M.: Aliphatic Polyesters: Abiotic and Biotic Degradation and Degradation Products. Vol. 157, pp. 1-26.
Hall, H. K. see Penelle, J.: Vol. 102, pp. 73-104.
Hamley, I. W.: Crystallization in Block Copolymers. Vol. 148, pp. 113-138.
Hammouda, B.: SANS from Homogeneous Polymer Mixtures: A Unified Overview. Vol. 106, pp. 87-134.
Han, M.J. and *Chang, J.Y.*: Polynucleotide Analogues. Vol. 153, pp. 1-36.
Harada, A.: Design and Construction of Supramolecular Architectures Consisting of Cyclodextrins and Polymers. Vol. 133, pp. 141-192.
Haralson, M. A. see Prokop, A.: Vol. 136, pp. 1-52.
Hassan, C.M. and *Peppas, N.A.*: Structure and Applications of Poly(vinyl alcohol) Hydrogels Produced by Conventional Crosslinking or by Freezing/Thawing Methods. Vol. 153, pp. 37-65.
Hawker, C. J. Dentritic and Hyperbranched Macromolecules – Precisely Controlled Macromolecular Architectures. Vol. 147, pp. 113-160.
Hawker, C. J. see Hedrick, J. L.: Vol. 141, pp. 1-44.
He, G. S. see Lin, T.-C.: Vol. 161, pp. 157-193.
Hedrick, J. L., Carter, K. R., Labadie, J. W., Miller, R. D., Volksen, W., Hawker, C. J., Yoon, D. Y., Russell, T. P., McGrath, J. E., Briber, R. M.: Nanoporous Polyimides. Vol. 141, pp. 1-44.
Hedrick, J. L., Labadie, J. W., Volksen, W. and *Hilborn, J. G.*: Nanoscopically Engineered Polyimides. Vol. 147, pp. 61-112.
Hedrick, J. L. see Hergenrother, P. M.: Vol. 117, pp. 67-110.
Hedrick, J. L. see Kiefer, J.: Vol. 147, pp. 161-247.
Hedrick, J.L. see McGrath, J. E.: Vol. 140, pp. 61-106.
Heinrich, G. and *Klüppel, M.*: Recent Advances in the Theory of Filler Networking in Elastomers. Vol. 160, pp. 1–44.
Heller, J.: Poly (Ortho Esters). Vol. 107, pp. 41-92.
Hemielec, A. A. see Hunkeler, D.: Vol. 112, pp. 115-134.
Hergenrother, P. M., Connell, J. W., Labadie, J. W. and *Hedrick, J. L.*: Poly(arylene ether)s Containing Heterocyclic Units. Vol. 117, pp. 67-110.
Hernández-Barajas, J. see Wandrey, C.: Vol. 145, pp. 123-182.
Hervet, H. see Léger, L.: Vol. 138, pp. 185-226.
Hilborn, J. G. see Hedrick, J. L.: Vol. 147, pp. 61-112.
Hilborn, J. G. see Kiefer, J.: Vol. 147, pp. 161-247.
Hiramatsu, N. see Matsushige, M.: Vol. 125, pp. 147-186.
Hirasa, O. see Suzuki, M.: Vol. 110, pp. 241-262.
Hirotsu, S.: Coexistence of Phases and the Nature of First-Order Transition in Poly-N-isopropylacrylamide Gels. Vol. 110, pp. 1-26.
Höcker, H. see Klee, D.: Vol. 149, pp. 1-57.
Hornsby, P.: Rheology, Compoundind and Processing of Filled Thermoplastics. Vol. 139, pp. 155-216.
Hui, C.-Y. see Creton, C.: Vol. 156, pp. 53-135
Hult, A., Johansson, M., Malmström, E.: Hyperbranched Polymers. Vol. 143, pp. 1-34.
Hunkeler, D., Candau, F., Pichot, C., Hemielec, A. E., Xie, T. Y., Barton, J., Vaskova, V., Guillot, J., Dimonie, M. V., Reichert, K. H.: Heterophase Polymerization: A Physical and Kinetic Comparision and Categorization. Vol. 112, pp. 115-134.

Hunkeler, D. see Prokop, A.: Vol. 136, pp. 1-52; 53-74.
Hunkeler, D see Wandrey, C.: Vol. 145, pp. 123-182.

Iatrou, H. see Hadjichristidis, N.: Vol. 142, pp. 71-128.
Ichikawa, T. see Yoshida, H.: Vol. 105, pp. 3-36.
Ihara, E. see Yasuda, H.: Vol. 133, pp. 53-102.
Ikada, Y. see Uyama, Y.: Vol. 137, pp. 1-40.
Ilavsky, M.: Effect on Phase Transition on Swelling and Mechanical Behavior of Synthetic Hydrogels. Vol. 109, pp. 173-206.
Imai, Y.: Rapid Synthesis of Polyimides from Nylon-Salt Monomers. Vol. 140, pp. 1-23.
Inomata, H. see Saito, S.: Vol. 106, pp. 207-232.
Inoue, S. see Sugimoto, H.: Vol. 146, pp. 39-120.
Irie, M.: Stimuli-Responsive Poly(N-isopropylacrylamide), Photo- and Chemical-Induced Phase Transitions. Vol. 110, pp. 49-66.
Ise, N. see Matsuoka, H.: Vol. 114, pp. 187-232.
Ito, K., Kawaguchi, S,:Poly(macronomers), Homo- and Copolymerization. Vol. 142, pp. 129-178.
Ivanov, A. E. see Zubov, V. P.: Vol. 104, pp. 135-176.

Jacob, S. and Kennedy, J.: Synthesis, Characterization and Properties of OCTA-ARM Polyisobutylene-Based Star Polymers. Vol. 146, pp. 1-38.
Jaffe, M., Chen, P., Choe, E.-W., Chung, T.-S. and *Makhija, S.*: High Performance Polymer Blends. Vol. 117, pp. 297-328.
Jancar, J.: Structure-Property Relationships in Thermoplastic Matrices. Vol. 139, pp. 1-66.
Jen, A. K-Y. see Kajzar, F.: Vol. 161, pp. 1-85.
Jerôme, R.: see Mecerreyes, D.: Vol. 147, pp. 1-60.
Jiang, M., Li, M., Xiang, M. and Zhou, H.: Interpolymer Complexation and Miscibility and Enhancement by Hydrogen Bonding. Vol. 146, pp. 121-194.
Jin, J.: see Shim, H.-K.: Vol. 158, pp. 191-241.
Jo, W. H. and Yang, J. S.: Molecular Simulation Approaches for Multiphase Polymer Systems. Vol. 156, pp. 1-52.
Johansson, M. see Hult, A.: Vol. 143, pp. 1-34.
Joos-Müller, B. see Funke, W.: Vol. 136, pp. 137-232.
Jou, D., Casas-Vazquez, J. and *Criado-Sancho, M.*: Thermodynamics of Polymer Solutions under Flow: Phase Separation and Polymer Degradation. Vol. 120, pp. 207-266.

Kaetsu, I.: Radiation Synthesis of Polymeric Materials for Biomedical and Biochemical Applications. Vol. 105, pp. 81-98.
Kaji, K. see Kanaya, T.: Vol. 154, pp. 87-141.
Kajzar, F., Lee, K.-S., Jen, A.K.-Y.: Polymeric Materials and their Orientation Techniques for Second-Order Nonlinear Optics.Vol. 161, pp. 1-85.
Kakimoto, M. see Gaw, K. O.: Vol. 140, pp. 107-136.
Kaminski, W. and *Arndt, M.*: Metallocenes for Polymer Catalysis. Vol. 127, pp. 143-187.
Kammer, H. W., Kressler, H. and *Kummerloewe, C.*: Phase Behavior of Polymer Blends - Effects of Thermodynamics and Rheology. Vol. 106, pp. 31-86.
Kanaya, T. and Kaji, K.: Dynamcis in the Glassy State and Near the Glass Transition of Amorphous Polymers as Studied by Neutron Scattering. Vol. 154, pp. 87-141.
Kandyrin, L. B. and *Kuleznev, V. N.*: The Dependence of Viscosity on the Composition of Concentrated Dispersions and the Free Volume Concept of Disperse Systems. Vol. 103, pp. 103-148.
Kaneko, M. see Ramaraj, R.: Vol. 123, pp. 215-242.
Kang, E. T., Neoh, K. G. and *Tan, K. L.*: X-Ray Photoelectron Spectroscopic Studies of Electroactive Polymers. Vol. 106, pp. 135-190.
Karlsson, S. see Söderqvist Lindblad, M.: Vol. 157, pp. 139–161.
Kato, K. see Uyama, Y.: Vol. 137, pp. 1-40.
Kawaguchi, S. see Ito, K.: Vol. 142, p 129-178.

Kazanskii, K. S. and *Dubrovskii, S. A.*: Chemistry and Physics of „Agricultural" Hydrogels. Vol. 104, pp. 97-134.

Kennedy, J. P. see Jacob, S.: Vol. 146, pp. 1-38.

Kennedy, J. P. see Majoros, I.: Vol. 112, pp. 1-113.

Khokhlov, A., Starodybtzev, S. and *Vasilevskaya, V.*: Conformational Transitions of Polymer Gels: Theory and Experiment. Vol. 109, pp. 121-172.

Kiefer, J., Hedrick J. L. and *Hiborn, J. G.*: Macroporous Thermosets by Chemically Induced Phase Separation. Vol. 147, pp. 161-247.

Kilian, H. G. and *Pieper, T.*: Packing of Chain Segments. A Method for Describing X-Ray Patterns of Crystalline, Liquid Crystalline and Non-Crystalline Polymers. Vol. 108, pp. 49-90.

Kim, J. see Quirk, R.P.: Vol. 153, pp. 67-162.

Kim, K.-S. see Lin, T.-C.: Vol. 161, pp. 157-193.

Kippelen, B. and Peyghambarian, N.: Photorefractive Polymers and their Applications. Vol. 161, pp. 87-156.

Kishore, K. and *Ganesh, K.*: Polymers Containing Disulfide, Tetrasulfide, Diselenide and Ditelluride Linkages in the Main Chain. Vol. 121, pp. 81-122.

Kitamaru, R.: Phase Structure of Polyethylene and Other Crystalline Polymers by Solid-State ^{13}C/MNR. Vol. 137, pp 41-102.

Klee, D. and *Höcker, H.*: Polymers for Biomedical Applications: Improvement of the Interface Compatibility. Vol. 149, pp. 1-57.

Klier, J. see Scranton, A. B.: Vol. 122, pp. 1-54.

Klüppel, M. see Heinrich, G.: Vol. 160, pp 1-44.

Kobayashi, S., Shoda, S. and *Uyama, H.*: Enzymatic Polymerization and Oligomerization. Vol. 121, pp. 1-30.

Köhler, W. and *Schäfer, R.*: Polymer Analysis by Thermal-Diffusion Forced Rayleigh Scattering. Vol. 151, pp. 1-59.

Koenig, J. L. see Andreis, M.: Vol. 124, pp. 191-238.

Koike, T.: Viscoelastic Behavior of Epoxy Resins Before Crosslinking. Vol. 148, pp. 139-188.

Kokufuta, E.: Novel Applications for Stimulus-Sensitive Polymer Gels in the Preparation of Functional Immobilized Biocatalysts. Vol. 110, pp. 157-178.

Konno, M. see Saito, S.: Vol. 109, pp. 207-232.

Kopecek, J. see Putnam, D.: Vol. 122, pp. 55-124.

Koßmehl, G. see Schopf, G.: Vol. 129, pp. 1-145.

Kozlov, E. see Prokop, A.: Vol. 160, pp. 119-174.

Kramer, E. J. see Creton, C.: Vol. 156, pp. 53-135.

Kremer, K. see Baschnagel, J.: Vol. 152, pp. 41-156.

Kressler, J. see Kammer, H. W.: Vol. 106, pp. 31-86.

Kricheldorf, H. R.: Liquid-Cristalline Polyimides. Vol. 141, pp. 83-188.

Krishnamoorti, R. see Giannelis, E.P.: Vol. 138, pp. 107-148.

Kirchhoff, R. A. and *Bruza, K. J.*: Polymers from Benzocyclobutenes. Vol. 117, pp. 1-66.

Kuchanov, S. I.: Modern Aspects of Quantitative Theory of Free-Radical Copolymerization. Vol. 103, pp. 1-102.

Kuchanov, S. I.: Principles of Quantitive Description of Chemical Structure of Synthetic Polymers. Vol. 152, p. 157-202.

Kudaibergennow, S.E.: Recent Advances in Studying of Synthetic Polyampholytes in Solutions. Vol. 144, pp. 115-198.

Kuleznev, V. N. see Kandyrin, L. B.: Vol. 103, pp. 103-148.

Kulichkhin, S. G. see Malkin, A. Y.: Vol. 101, pp. 217-258.

Kulicke, W.-M. see Grigorescu, G.: Vol. 152, p. 1-40.

Kumar, M.N.V.R., Kumar, N., Domb, A.J. and Arora, M.: Pharmaceutical Polyme-ric Controlled Drug Delivery Systems. Vol. 160, pp. 45-118.

Kumar, N. see Kumar M.N.V.R.: Vol. 160, pp. 45-118.

Kummerloewe, C. see Kammer, H. W.: Vol. 106, pp. 31-86.

Kuznetsova, N. P. see Samsonov, G. V.: Vol. 104, pp. 1-50. Labadie, J. W. see Hergenrother, P. M.: Vol. 117, pp. 67-110.

Labadie, J. W. see Hedrick, J. L.: Vol. 141, pp. 1-44.
Labadie, J. W. see Hedrick, J. L.: Vol. 147, pp. 61-112.
Lamparski, H. G. see O´Brien, D. F.: Vol. 126, pp. 53-84.
Laschewsky, A.: Molecular Concepts, Self-Organisation and Properties of Polysoaps. Vol. 124, pp. 1-86.
Laso, M. see Leontidis, E.: Vol. 116, pp. 283-318.
Lazár, M. and RychlΩ, R.: Oxidation of Hydrocarbon Polymers. Vol. 102, pp. 189-222.
Lechowicz, J. see Galina, H.: Vol. 137, pp. 135-172.
Léger, L., Raphaël, E., Hervet, H.: Surface-Anchored Polymer Chains: Their Role in Adhesion and Friction. Vol. 138, pp. 185-226.
Lenz, R. W.: Biodegradable Polymers. Vol. 107, pp. 1-40.
Leontidis, E., de Pablo, J. J., Laso, M. and *Suter, U. W.*: A Critical Evaluation of Novel Algorithms for the Off-Lattice Monte Carlo Simulation of Condensed Polymer Phases. Vol. 116, pp. 283-318.
Lee, B. see Quirk, R.P: Vol. 153, pp. 67-162.
Lee, K.-S. see Kajzar, F.: Vol. 161, pp. 1-85.
Lee, Y. see Quirk, R.P: Vol. 153, pp. 67-162.
Lesec, J. see Viovy, J.-L.: Vol. 114, pp. 1-42.
Li, M. see Jiang, M.: Vol. 146, pp. 121-194.
Liang, G. L. see Sumpter, B. G.: Vol. 116, pp. 27-72.
Lienert, K.-W.: Poly(ester-imide)s for Industrial Use. Vol. 141, pp. 45-82.
Lin, J. and *Sherrington, D. C.*: Recent Developments in the Synthesis, Thermostability and Liquid Crystal Properties of Aromatic Polyamides. Vol. 111, pp. 177-220.
Lin, T.-C., Chung, S.-J., Kim, K.-S., Wang, X., He, G. S., Swiatkiewicz, J., Pudavar, H. E. and Prasad, P. N.: Organics and Polymers with High Two-Photon Activities and their Applications. Vol. 161, pp. 157-193.
Liu, Y. see Söderqvist Lindblad, M.: Vol. 157, pp. 139-161
López Cabarcos, E. see Baltá-Calleja, F. J.: Vol. 108, pp. 1-48.

Majoros, I., Nagy, A. and *Kennedy, J. P.*: Conventional and Living Carbocationic Polymerizations United. I. A Comprehensive Model and New Diagnostic Method to Probe the Mechanism of Homopolymerizations. Vol. 112, pp. 1-113.
Makhija, S. see Jaffe, M.: Vol. 117, pp. 297-328.
Malmström, E. see Hult, A.: Vol. 143, pp. 1-34.
Malkin, A. Y. and *Kulichkhin, S. G.*: Rheokinetics of Curing. Vol. 101, pp. 217-258.
Maniar, M. see Domb, A. J.: Vol. 107, pp. 93-142.
Manias, E., see Giannelis, E.P.: Vol. 138, pp. 107-148.
Mashima, K., Nakayama, Y. and *Nakamura, A.*: Recent Trends in Polymerization of a-Olefins Catalyzed by Organometallic Complexes of Early Transition Metals. Vol. 133, pp. 1-52.
Mathew, D. see Reghunadhan Nair, C.P.: Vol. 155, pp. 1-99.
Matsumoto, A.: Free-Radical Crosslinking Polymerization and Copolymerization of Multivinyl Compounds. Vol. 123, pp. 41-80.
Matsumoto, A. see Otsu, T.: Vol. 136, pp. 75-138.
Matsuoka, H. and *Ise, N.*: Small-Angle and Ultra-Small Angle Scattering Study of the Ordered Structure in Polyelectrolyte Solutions and Colloidal Dispersions. Vol. 114, pp. 187-232.
Matsushige, K., Hiramatsu, N. and *Okabe, H.*: Ultrasonic Spectroscopy for Polymeric Materials. Vol. 125, pp. 147-186.
Mattice, W. L. see Rehahn, M.: Vol. 131/132, pp. 1-475.
Mattice, W. L. see Baschnagel, J.: Vol. 152, p. 41-156.
Mays, W. see Xu, Z.: Vol. 120, pp. 1-50.
Mays, J.W. see Pitsikalis, M.: Vol.135, pp. 1-138.
McGrath, J. E. see Hedrick, J. L.: Vol. 141, pp. 1-44.
McGrath, J. E., Dunson, D. L., Hedrick, J. L.: Synthesis and Characterization of Segmented Polyimide-Polyorganosiloxane Copolymers. Vol. 140, pp. 61-106.
McLeish, T.C. B., Milner, S. T.: Entangled Dynamics and Melt Flow of Branched Polymers. Vol. 143, pp. 195-256.

Mecerreyes, D., Dubois, P. and *Jerôme, R.*: Novel Macromolecular Architectures Based on Aliphatic Polyesters: Relevance of the „Coordination-Insertion" Ring-Opening Polymerization. Vol. 147, pp. 1 -60.

Mecham, S. J. see McGrath, J. E.: Vol. 140, pp. 61-106.

Mikos, A. G. see Thomson, R. C.: Vol. 122, pp. 245-274.

Milner, S. T. see McLeish, T. C. B.: Vol. 143, pp. 195-256.

Mison, P. and Sillion, B.: Thermosetting Oligomers Containing Maleimides and Nadiimides End-Groups. Vol. 140, pp. 137-180.

Miyasaka, K.: PVA-Iodine Complexes: Formation, Structure and Properties. Vol. 108. pp. 91-130.

Miller, R. D. see Hedrick, J. L.: Vol. 141, pp. 1-44.

Monnerie, L. see Bahar, I.: Vol. 116, pp. 145-206.

Morishima, Y.: Photoinduced Electron Transfer in Amphiphilic Polyelectrolyte Systems. Vol. 104, pp. 51-96.

Morton M. see Quirk, R.P: Vol. 153, pp. 67-162

Mours, M. see Winter, H. H.: Vol. 134, pp. 165-234.

Müllen, K. see Scherf, U.: Vol. 123, pp. 1-40.

Müller-Plathe, F. see Gusev, A. A.: Vol. 116, pp. 207-248.

Müller-Plathe, F. see Baschnagel, J.: Vol. 152, p. 41-156.

Mukerherjee, A. see Biswas, M.: Vol. 115, pp. 89-124.

Murat, M. see Baschnagel, J.: Vol. 152, p. 41-156.

Mylnikov, V.: Photoconducting Polymers. Vol. 115, pp. 1-88.

Nagy, A. see Majoros, I.: Vol. 112, pp. 1-11.

Nakamura, A. see Mashima, K.: Vol. 133, pp. 1-52.

Nakayama, Y. see Mashima, K.: Vol. 133, pp. 1-52.

Narasinham, B., Peppas, N. A.: The Physics of Polymer Dissolution: Modeling Approaches and Experimental Behavior. Vol. 128, pp. 157-208.

Nechaev, S. see Grosberg, A.: Vol. 106, pp. 1-30.

Neoh, K. G. see Kang, E. T.: Vol. 106, pp. 135-190.

Newman, S. M. see Anseth, K. S.: Vol. 122, pp. 177-218.

Nijenhuis, K. te: Thermoreversible Networks. Vol. 130, pp. 1-252.

Ninan, K.N. see Reghunadhan Nair, C. P.: Vol. 155, pp. 1-99.

Noid, D. W. see Otaigbe, J.U.: Vol. 154, pp. 1-86.

Noid, D. W. see Sumpter, B. G.: Vol. 116, pp. 27-72.

Novac, B. see Grubbs, R.: Vol. 102, pp. 47-72.

Novikov, V. V. see Privalko, V. P.: Vol. 119, pp. 31-78.

O'Brien, D. F., Armitage, B. A., Bennett, D. E. and *Lamparski, H. G.*: Polymerization and Domain Formation in Lipid Assemblies. Vol. 126, pp. 53-84.

Ogasawara, M.: Application of Pulse Radiolysis to the Study of Polymers and Polymerizations. Vol.105, pp. 37-80.

Okabe, H. see Matsushige, K.: Vol. 125, pp. 147-186.

Okada, M.: Ring-Opening Polymerization of Bicyclic and Spiro Compounds. Reactivities and Polymerization Mechanisms. Vol. 102, pp. 1-46.

Okano, T.: Molecular Design of Temperature-Responsive Polymers as Intelligent Materials. Vol. 110, pp. 179-198.

Okay, O. see Funke, W.: Vol. 136, pp. 137-232.

Onuki, A.: Theory of Phase Transition in Polymer Gels. Vol. 109, pp. 63-120.

Osad'ko, I.S.: Selective Spectroscopy of Chromophore Doped Polymers and Glasses. Vol. 114, pp. 123-186.

Otaigbe, J. U., Barnes, M. D., Fukui, K., Sumpter, B. G., Noid, D. W.: Generation, Characterization, and Modeling of Polymer Micro- and Nano-Particles. Vol. 154, pp. 1-86.

Otsu, T., Matsumoto, A.: Controlled Synthesis of Polymers Using the Iniferter Technique: Developments in Living Radical Polymerization. Vol. 136, pp. 75-138.

de Pablo, J. J. see Leontidis, E.: Vol. 116, pp. 283-318.

Padias, A. B. see Penelle, J.: Vol. 102, pp. 73-104.

Pascault, J.-P. see Williams, R. J. J.: Vol. 128, pp. 95-156.

Pasch, H.: Analysis of Complex Polymers by Interaction Chromatography. Vol. 128, pp. 1-46.

Pasch, H.: Hyphenated Techniques in Liquid Chromatography of Polymers. Vol. 150, pp. 1-66.

Paul, W. see Baschnagel, J.: Vol. 152, p. 41-156.

Penczek, P. see Batog, A. E.: Vol. 144, pp. 49-114.

Penelle, J., Hall, H. K., Padias, A. B. and *Tanaka, H.*: Captodative Olefins in Polymer Chemistry. Vol. 102, pp. 73-104.

Peppas, N. A. see Bell, C. L.: Vol. 122, pp. 125-176.

Peppas, N.A. see Hassan, C.M.: Vol. 153, pp. 37-65

Peppas, N. A. see Narasimhan, B.: Vol. 128, pp. 157-208.

Pet'ko, I. P. see Batog, A. E.: Vol. 144, pp. 49-114.

Pheyghambarian, N. see Kippelen, B.: Vol. 161, pp. 87-156.

Pichot, C. see Hunkeler, D.: Vol. 112, pp. 115-134.

Pieper, T. see Kilian, H. G.: Vol. 108, pp. 49-90.

Pispas, S. see Pitsikalis, M.: Vol. 135, pp. 1-138.

Pispas, S. see Hadjichristidis: Vol. 142, pp. 71-128.

Pitsikalis, M., Pispas, S., Mays, J. W., Hadjichristidis, N.: Nonlinear Block Copolymer Architectures. Vol. 135, pp. 1-138.

Pitsikalis, M. see Hadjichristidis: Vol. 142, pp. 71-128.

Pötschke, D. see Dingenouts, N.: Vol 144, pp. 1-48.

Pokrovskii, V. N.: The Mesoscopic Theory of the Slow Relaxation of Linear Macromolecules. Vol. 154, pp. 143-219.

Pospíšil, J.: Functionalized Oligomers and Polymers as Stabilizers for Conventional Polymers. Vol. 101, pp. 65-168.

Pospíšil, J.: Aromatic and Heterocyclic Amines in Polymer Stabilization. Vol. 124, pp. 87-190.

Powers, A. C. see Prokop, A.: Vol. 136, pp. 53-74.

Prasad, P. N. see Lin, T.-C.: Vol. 161, pp. 157-193.

Priddy, D. B.: Recent Advances in Styrene Polymerization. Vol. 111, pp. 67-114.

Priddy, D. B.: Thermal Discoloration Chemistry of Styrene-co-Acrylonitrile. Vol. 121, pp. 123-154.

Privalko, V. P. and *Novikov, V. V.*: Model Treatments of the Heat Conductivity of Heterogeneous Polymers. Vol. 119, pp 31-78.

Prokop, A., Hunkeler, D., Powers, A. C., Whitesell, R. R., Wang, T. G.: Water Soluble Polymers for Immunoisolation II: Evaluation of Multicomponent Microencapsulation Systems. Vol. 136, pp. 53-74.

Prokop, A., Hunkeler, D., DiMari, S., Haralson, M. A., Wang, T. G.: Water Soluble Polymers for Immunoisolation I: Complex Coacervation and Cytotoxicity. Vol. 136, pp. 1-52.

Prokop, A., Kozlov, E., Carlesso, G. and Davidsen, J.M.: Hydrogel-Based Colloidal Polymeric System for Protein and Drug Delivery: Physical and Chemical Characterization, Permeability Control and Applications. Vol. 160, pp. 119-174.

Pudavar, H. E. see Lin, T.-C.: Vol. 161, pp. 157-193.

Pukánszky, B. and *Fekete, E.*: Adhesion and Surface Modification. Vol. 139, pp. 109-154.

Putnam, D. and *Kopecek, J.*: Polymer Conjugates with Anticancer Acitivity. Vol. 122, pp. 55- 124.

Quirk, R.P. and Yoo, T., Lee, Y., M., Kim, J. and Lee, B.: Applications of 1,1-Diphenylethylene Chemistry in Anionic Synthesis of Polymers with Controlled Structures. Vol. 153, pp. 67-162.

Ramaraj, R. and *Kaneko, M.*: Metal Complex in Polymer Membrane as a Model for Photosynthetic Oxygen Evolving Center. Vol. 123, pp. 215-242.

Rangarajan, B. see Scranton, A. B.: Vol. 122, pp. 1-54.

Ranucci, E. see Söderqvist Lindblad, M.: Vol. 157, pp. 139–161.

Raphaël, E. see Léger, L.: Vol. 138, pp. 185-226.

Reddinger, J. L. and *Reynolds, J. R.*: Molecular Engineering of π-Conjugated Polymers. Vol. 145, pp. 57-122.

Reghunadhan Nair, C.P., Mathew, D. and *Ninan, K.N.,* : Cyanate Ester Resins, Recent Developments. Vol. 155, pp. 1-99.

Reichert, K. H. see Hunkeler, D.: Vol. 112, pp. 115-134.

Rehahn, M., Mattice, W. L., Suter, U. W.: Rotational Isomeric State Models in Macromolecular Systems. Vol. 131/132, pp. 1-475.

Reynolds, J.R. see Reddinger, J. L.: Vol. 145, pp. 57-122.

Richter, D. see Ewen, B.: Vol. 134, pp.1-130.

Risse, W. see Grubbs, R.: Vol. 102, pp. 47-72.

Rivas, B. L. and *Geckeler, K. E.*: Synthesis and Metal Complexation of Poly(ethyleneimine) and Derivatives. Vol. 102, pp. 171-188.

Robin, J. J. see Boutevin, B.: Vol. 102, pp. 105-132.

Roe, R.-J.: MD Simulation Study of Glass Transition and Short Time Dynamics in Polymer Liquids. Vol. 116, pp. 111-114.

Roovers, J., Comanita, B.: Dendrimers and Dendrimer-Polymer Hybrids. Vol. 142, pp 179-228.

Rothon, R. N.: Mineral Fillers in Thermoplastics: Filler Manufacture and Characterisation. Vol. 139, pp. 67-108.

Rozenberg, B. A. see Williams, R. J. J.: Vol. 128, pp. 95-156.

Ruckenstein, E.: Concentrated Emulsion Polymerization. Vol. 127, pp. 1-58.

Rusanov, A. L.: Novel Bis (Naphtalic Anhydrides) and Their Polyheteroarylenes with Improved Processability. Vol. 111, pp. 115-176.

Russel, T. P. see Hedrick, J. L.: Vol. 141, pp. 1-44.

Rychlý, J. see Lazár, M.: Vol. 102, pp. 189-222.

Ryner, M. see Stridsberg, K. M.: Vol. 157, pp. 27–51.

Ryzhov, V. A. see Bershtein, V. A.: Vol. 114, pp. 43-122.

Sabsai, O. Y. see Barshtein, G. R.: Vol. 101, pp. 1-28.

Saburov, V. V. see Zubov, V. P.: Vol. 104, pp. 135-176.

Saito, S., Konno, M. and *Inomata, H.*: Volume Phase Transition of N-Alkylacrylamide Gels. Vol. 109, pp. 207-232.

Samsonov, G. V. and *Kuznetsova, N. P.*: Crosslinked Polyelectrolytes in Biology. Vol. 104, pp. 1-50.

Santa Cruz, C. see Baltá-Calleja, F. J.: Vol. 108, pp. 1-48.

Santos, S. see Baschnagel, J.: Vol. 152, p. 41-156.

Sato, T. and *Teramoto, A.*: Concentrated Solutions of Liquid-Christalline Polymers. Vol. 126, pp. 85-162.

Schäfer R. see Köhler, W.: Vol. 151, pp. 1-59.

Scherf, U. and *Müllen, K.*: The Synthesis of Ladder Polymers. Vol. 123, pp. 1-40.

Schmidt, M. see Förster, S.: Vol. 120, pp. 51-134.

Schopf, G. and *Koßmehl, G.*: Polythiophenes - Electrically Conductive Polymers. Vol. 129, pp. 1-145.

Schweizer, K. S.: Prism Theory of the Structure, Thermodynamics, and Phase Transitions of Polymer Liquids and Alloys. Vol. 116, pp. 319-378.

Scranton, A. B., Rangarajan, B. and *Klier, J.*: Biomedical Applications of Polyelectrolytes. Vol. 122, pp. 1-54.

Sefton, M. V. and *Stevenson, W. T. K.*: Microencapsulation of Live Animal Cells Using Polycrylates. Vol.107, pp. 143-198.

Shamanin, V. V.: Bases of the Axiomatic Theory of Addition Polymerization. Vol. 112, pp. 135-180.

Sheiko, S. S.: Imaging of Polymers Using Scanning Force Microscopy: From Superstructures to Individual Molecules. Vol. 151, pp. 61-174.

Sherrington, D. C. see Cameron, N. R. , Vol. 126, pp. 163-214.

Sherrington, D. C. see Lin, J.: Vol. 111, pp. 177-220.

Sherrington, D. C. see Steinke, J.: Vol. 123, pp. 81-126.

Shibayama, M. see Tanaka, T.: Vol. 109, pp. 1-62.

Shiga, T.: Deformation and Viscoelastic Behavior of Polymer Gels in Electric Fields. Vol. 134, pp. 131-164.

Shim, H.-K., Jin, J.: Light-Emitting Characteristics of Conjugated Polymers. Vol. 158, pp. 191-241.

Shoda, S. see Kobayashi, S.: Vol. 121, pp. 1-30.

Siegel, R. A.: Hydrophobic Weak Polyelectrolyte Gels: Studies of Swelling Equilibria and Kinetics. Vol. 109, pp. 233-268.

Silvestre, F. see Calmon-Decriaud, A.: Vol. 207, pp. 207-226.

Sillion, B. see Mison, P.: Vol. 140, pp. 137-180.

Singh, R. P. see Sivaram, S.: Vol. 101, pp. 169-216.

Sinha Ray, S. see Biswas, M: Vol. 155, pp. 167-221.

Sivaram, S. and *Singh, R. P.*: Degradation and Stabilization of Ethylene-Propylene Copolymers and Their Blends: A Critical Review. Vol. 101, pp. 169-216.

Söderqvist Lindblad, M., Liu, Y., Albertsson, A.-C., Ranucci, E., Karlsson, S.: Polymer from Renewable Resources. Vol. 157, pp. 139–161

Starodybtzev, S. see Khokhlov, A.: Vol. 109, pp. 121-172.

Stegeman, G. I.: see Canva, M.: Vol. 158, pp. 87-121.

Steinke, J., Sherrington, D. C. and *Dunkin, I. R.*: Imprinting of Synthetic Polymers Using Molecular Templates. Vol. 123, pp. 81-126.

Stenzenberger, H. D.: Addition Polyimides. Vol. 117, pp. 165-220.

Stevenson, W. T. K. see Sefton, M. V.: Vol. 107, pp. 143-198.

Stridsberg, K. M., Ryner, M., Albertsson, A.-C.: Controlled Ring-Opening Polymerization: Polymers with Designed Macromoleculars Architecture. Vol. 157, pp. 27–51.

Suematsu, K.: Recent Progress of Gel Theory: Ring, Excluded Volume, and Dimension. Vol. 156, pp. 136-214.

Sumpter, B. G., Noid, D. W., Liang, G. L. and *Wunderlich, B.*: Atomistic Dynamics of Macromolecular Crystals. Vol. 116, pp. 27-72.

Sumpter, B. G. see Otaigbe, J.U.: Vol. 154, pp. 1-86.

Sugimoto, H. and *Inoue, S.*: Polymerization by Metalloporphyrin and Related Complexes. Vol. 146, pp. 39-120.

Suter, U. W. see Gusev, A. A.: Vol. 116, pp. 207-248.

Suter, U. W. see Leontidis, E.: Vol. 116, pp. 283-318.

Suter, U. W. see Rehahn, M.: Vol. 131/132, pp. 1-475.

Suter, U. W. see Baschnagel, J.: Vol. 152, p. 41-156.

Suzuki, A.: Phase Transition in Gels of Sub-Millimeter Size Induced by Interaction with Stimuli. Vol. 110, pp. 199-240.

Suzuki, A. and *Hirasa, O.*: An Approach to Artifical Muscle by Polymer Gels due to Micro-Phase Separation. Vol. 110, pp. 241-262.

Swiatkiewicz, J. see Lin, T.-C.: Vol. 161, pp. 157-193.

Tagawa, S.: Radiation Effects on Ion Beams on Polymers. Vol. 105, pp. 99-116.

Tan, K. L. see Kang, E. T.: Vol. 106, pp. 135-190.

Tanaka, H. and *Shibayama, M.*: Phase Transition and Related Phenomena of Polymer Gels. Vol. 109, pp. 1-62.

Tanaka, T. see Penelle, J.: Vol. 102, pp. 73-104.

Tauer, K. see Guyot, A.: Vol. 111, pp. 43-66.

Teramoto, A. see Sato, T.: Vol. 126, pp. 85-162.

Terent´eva, J. P. and *Fridman, M. L.*: Compositions Based on Aminoresins. Vol. 101, pp. 29-64.

Theodorou, D. N. see Dodd, L. R.: Vol. 116, pp. 249-282.

Thomson, R. C., Wake, M. C., Yaszemski, M. J. and *Mikos, A. G.*: Biodegradable Polymer Scaffolds to Regenerate Organs. Vol. 122, pp. 245-274.

Tokita, M.: Friction Between Polymer Networks of Gels and Solvent. Vol. 110, pp. 27-48.

Tries, V. see Baschnagel, J:. Vol. 152, p. 41-156.

Tsuruta, T.: Contemporary Topics in Polymeric Materials for Biomedical Applications. Vol. 126, pp. 1-52.

Uyama, H. see Kobayashi, S.: Vol. 121, pp. 1-30.
Uyama, Y: Surface Modification of Polymers by Grafting. Vol. 137, pp. 1-40.

Varma, I. K. see Albertsson, A.-C.: Vol. 157, pp. 99-138.
Vasilevskaya, V. see Khokhlov, A.: Vol. 109, pp. 121-172.
Vaskova, V. see Hunkeler, D.: Vol.:112, pp. 115-134.
Verdugo, P.: Polymer Gel Phase Transition in Condensation-Decondensation of Secretory Products. Vol. 110, pp. 145-156.
Vettegren, V. I.: see Bronnikov, S. V.: Vol. 125, pp. 103-146.
Viovy, J.-L. and *Lesec, J.:* Separation of Macromolecules in Gels: Permeation Chromatography and Electrophoresis. Vol. 114, pp. 1-42.
Vlahos, C. see Hadjichristidis, N.: Vol. 142, pp. 71-128.
Volksen, W.: Condensation Polyimides: Synthesis, Solution Behavior, and Imidization Characteristics. Vol. 117, pp. 111-164.
Volksen, W. see Hedrick, J. L.: Vol. 141, pp. 1-44.
Volksen, W. see Hedrick, J. L.: Vol. 147, pp. 61-112.

Wake, M. C. see Thomson, R. C.: Vol. 122, pp. 245-274.
Wandrey C., Hernández-Barajas, J. and *Hunkeler, D.:* Diallyldimethylammonium Chloride and its Polymers. Vol. 145, pp. 123-182.
Wang, K. L. see Cussler, E. L.: Vol. 110, pp. 67-80.
Wang, S.-Q.: Molecular Transitions and Dynamics at Polymer/Wall Interfaces: Origins of Flow Instabilities and Wall Slip. Vol. 138, pp. 227-276.
Wang, T. G. see Prokop, A.: Vol. 136, pp.1-52; 53-74.
Wang, X. see Lin, T.-C.: Vol. 161, pp. 157-193.
Whitesell, R. R. see Prokop, A.: Vol. 136, pp. 53-74.
Williams, R. J. J., Rozenberg, B. A., Pascault, J.-P.: Reaction Induced Phase Separation in Modified Thermosetting Polymers. Vol. 128, pp. 95-156.
Winter, H. H., Mours, M.: Rheology of Polymers Near Liquid-Solid Transitions. Vol. 134, pp. 165-234.
Wu, C.: Laser Light Scattering Characterization of Special Intractable Macromolecules in Solution. Vol 137, pp. 103-134.
Wunderlich, B. see Sumpter, B. G.: Vol. 116, pp. 27-72.

Xiang, M. see Jiang, M.: Vol. 146, pp. 121-194.
Xie, T. Y. see Hunkeler, D.: Vol. 112, pp. 115-134.
Xu, Z., Hadjichristidis, N., Fetters, L. J. and *Mays, J. W.:* Structure/Chain-Flexibility Relationships of Polymers. Vol. 120, pp. 1-50.

Yagci, Y. and *Endo, T.:* N-Benzyl and N-Alkoxy Pyridium Salts as Thermal and Photochemical Initiators for Cationic Polymerization. Vol. 127, pp. 59-86.
Yannas, I. V.: Tissue Regeneration Templates Based on Collagen-Glycosaminoglycan Copolymers. Vol. 122, pp. 219-244.
Yang, J. S. see Jo, W. H.: Vol. 156, pp. 1-52.
Yamaoka, H.: Polymer Materials for Fusion Reactors. Vol. 105, pp. 117-144.
Yasuda, H. and *Ihara, E.:* Rare Earth Metal-Initiated Living Polymerizations of Polar and Nonpolar Monomers. Vol. 133, pp. 53-102.
Yaszemski, M. J. see Thomson, R. C.: Vol. 122, pp. 245-274.
Yoo, T. see Quirk, R.P.: Vol. 153, pp. 67-162.
Yoon, D. Y. see Hedrick, J. L.: Vol. 141, pp. 1-44.
Yoshida, H. and *Ichikawa, T.:* Electron Spin Studies of Free Radicals in Irradiated Polymers. Vol. 105, pp. 3-36.

Zhou, H. see Jiang, M.: Vol. 146, pp. 121-194.
Zubov, V. P., Ivanov, A. E. and *Saburov, V. V.:* Polymer-Coated Adsorbents for the Separation of Biopolymers and Particles. Vol. 104, pp. 135-176.

Subject Index

2,4-Toluene diisocyanate 17
2BNCM 145
2-Phenyltetracyanobutadienyl 16
4,4'-Diisocyanato-3,3'-dimethoxyphenyl 17
4'Dimethylamino-*N*-methyl-4-stilbazolium
  tosylate (DAST) 79

Absorption coefficient 96
Acceptor 108, 124
Adamantyl group 33
Aggregation 185
Alkoxysilane-functinalized NLO
  chromophore 44
Amorphous organic materials 120
Amorphous polycarbonate (APC) 33
Amplitude of modulation voltage 13
Anisotropy 101
AODCST 141
α-Relaxation mode 59
Aromatic polyimides 21
ATOP 140
Attenuated total reflection 60

Babinet-Soleil-Bravais (BSB)
  compensator 12
Bässler formalism 136
$BaTiO_3$ 120
BBP 145
Beam fanning 150
$Bi_{12}SiO_{20}$ 120
Birefringence 98
Bonding direction 27, 47
bondlength alternation 112
Bond-oder alternation 13

C_{60} 143
Carbazole moiety 143
Carbazole trimers 145

Carboxylate-containing dendrons 40
Carriers 123
Causality 97
CdTe 120
Centrosymmetric materials 6
Cerenkov-type SHG generation 20, 21, 74
Channel waveguide 45
Charge transfer 110
Charge-transfer complex 143
Charge-transfer molecules 6, 7, 65
Chromophore aggregation 16
Chromophore loading level 33, 38
Chromophore orientation techniques 50
Chromophore-chromophore electrostatic
  interaction 33, 38, 40, 43
Chromophores 118
Cis-trans isomerization 19, 72
Complex refractive index 96, 130
Confocal fluorescence microscopy 181
Constitutive equations 93
Contact poling 50
Continuity equation 125
Conversion efficiency 20, 76
Convolution product 96
Corona poling 50, 51, 63
Counter-propagating beams 75
Coupled-wave theory 147
Crosslinkable NLO dendrimers 32, 38, 40
Crosslinked NLO polymers 24, 26
Crystals 120
Cyanine limit 112
Cytotoxic agent 182

DEANST 140
Decay 40
Degeneracy factor 103
Dendrimers 28
DHADC-MPN 140

Diagonal disorder 136
Dielectric constant dispersion 10
Dielectric tensor 107
Dielectrics 94
Diethyl azodicarboxylate (DEAD) 21
Differential scanning calorimetry
  (DSC) 17
Diffraction efficiency 129
Diffusion field 126
Dipolar component 137
Dipolar dendrimers 29
Dipolar transition moment 54
Dipole moment 98, 108
Director cosines 115
Disorder formalism 136
Dispersion free hyperpolarizability 111
Dispersion relation 95
DMNPAA 140
Donor 108, 124
DOP 146
Doxorubicin (Dox) 182
DPANST 146
DRDCTA 146
δ-Relaxation mode 59
Dye-attached-sol-gel system 44
Dynamic holographic storage 149
Dynamic range modulation 130
Dynamic response 149

ECZ 145
Effective photorefractive trap density 126
Electric field-induced second harmonic
  generation (EFISH) 6, 7, 8, 16
Electric permittivity 94
Electro-active chromophores 140
Electron beam poling 50
Eectronic affinity 108
Electro-optic coefficient 11, 14, 78, 106
Electro-optic effect 9, 105
Electro-optic modulation 77
Electro-optic tensor elements 107
Energy transfer 123
Euler angles 116
Excited-state absorption 163, 173
Expansion coefficient 5
Extended polyene 14
External field strength 10
Extinction ratio 49
Extraordinary index 98

Femtosecond laser pulse 79
Figure of merit 131
First hyperpolarizability 109
Four-wave mixing 128, 172

Four-wave mixing geometry 63
Frequency doubling 19, 73
Frequency tuning 76
Fringe visibility 121
Full permutation symmetry 103
Functionalized polymers 147

GaAs 120
Gain coefficient 130, 131
Gas-phase model 13
Glass transition temperature 114
Grating spacing 121
Grating vector 135
Group theory 104
Guest-host approach 133
Guest-host type polymers 13, 39, 44, 60

Half-wave voltage 47, 78
Harmonic light scattering (HLS) 8, 30
Heterocyclic electron acceptors 14
Hilbert pair 97
Holographic information 120
Holographic recording 132
Homogeneous 94
Hopping site manifold 136
Hyper-Rayleigh scattering 8, 28

Image processing 149
Image recognition 149
Impermeability tensor 107
Index ellipsoid 106
Induced dipole moment 116
InP 120
Intensity 95
Intensity distribution 121
Intensity modulation response 48
Interaction energy 116
Interference pattern 136
Intermolecular electrostatic
  interactions 33
Intermolecular hopping 136
Intrinsic permutation symmetry 103
Inversion symmetry 101
Isotropic 94
Isotropic polymers 54

Kerr effect 105, 147
Kleinman symmetry 103
Kleinmen conditions 9, 65
$KnbO_3$ 120
Kogelnik 129
Kohlrausch-Williams-Watt (KWW)
  function 70
Kramers-Kronig relations 69, 97

Kukhtarev model 120
Laboratory frame 114
Langevin functions 117
Langmuir-Blodgett (LB) films 19
Lattice hardening 28, 32, 47
Legendre polynomials 67
Light-induced depoling 71
Light-induced grating 129
Limiting space-charge field 126
$LiNbO_3$ 120
Linear 94
Linear electro-optic effect 10
Liquid crystals 147
Lithium niobate 14, 109
Local field correction 115
Local field factor 7
Long-term stability 15, 26, 70
Lorentz oscillator 96
Luteinizing hormone-releasing hormone
   (LH-RH) 183

Mach-Zehnder modulator 45, 77
Macroscopic susceptibility 5
Maker fringe technique 25
Maxwell-Boltzmann distribution 116
Methansulfonylacetonitrile 25
Mitsunobu reaction 24
Mobility 123
Modal phase matching 76
Modulation ellipsometry technique 11
Molar extinction coefficient 8
Molecular axis 94
Molecular first hyperpolarizability (β) 5, 6
Molecular frame
Molecular polarizability (α) 5
Molecular second hyperpolarizability (γ) 5, 8
Multi-branched chromophore 165
Multifunctional 120
Multiple-chromophore-containing
   crosslinkable NLO dendrimers 37
Multiple-dendron-modified NLO
   chromophores 34

N-(4-Nitrophenyl)-(S)-prolinol (NPP) 44
Nematic liquid crystals 54
Noncentrosymmetric molecules 6
Noncentrosymmetry 101
Nondispersive 94
Nonlinear chromophores 32
Nonlinear optical response 100
Nonlinear optics 100
Non-linear transmission 168
Novelty filtering 152
Number density 7, 15

Octupolar ordering 29
Onsager 136
Operating voltage 49
Optical correlation 149
Optical correlator 150
Optical loss 44, 49
Optical parametric oscillators 76
Optical poling 50, 62
Optical power limiting 159, 174, 186, 189
Optical susceptibility 94
Order parameter 54, 57, 68
Ordinary index 98
Organic single crystal 131
Organic-inorganic hybrid materials 44
Orientational birefringence 123, 138
Orientational distribution function 116
Orientational ensemble 115
Orientational photorefractive effect 134
Orientational photorefractivity 138
Oriented gas model 114
Oscillator strength 110
Oxidation 124

Parallel locking structure 28
Pattern recognition 149
PBMA 144
PC 144
PDCST 141
Perfluorodendrons 35
Phase conjugation 149
Phase doubling 149
Phase separation 39
Photoassisted poling 50, 60, 64
Photo-bleaching 19, 180
Photoconductors 136
Photo-degradation 185
Photodynamic therapy 184
Photofrin® 184
Photogeneration 123, 133
Photogeneration efficiency 136
Photo-ionization cross section 125
Photoisomerization 61
Photo-polymerization 180
Photorefractive effect 120
Photorefractivity 120
Photosensitizer 184
Phototermal poling 50, 52
plasticizer 140, 145
PMMA 144
Pockels effect 9, 105
Pockels tensor components 11
Polar orientation 62
Polarizability anisotropy 138
Polarization 5, 94

Poled hybrid films 47
Poled-polymer approach 114
Poling 114
Poling efficiency 32, 33, 58
Poling field 116
Poling process 17
Poly(methyl methacryate) (PMMA) 14, 56
Poly[(phenyl methacryate)-co-formalde-
    hyde] (PPIF) 26
Polyacrylate 56
Polycarbonate (PC) 14
Polyenes 112
Polyetherimide (PEI) 21
Polyimide 15, 21
Polymer composites 120
polymer dispersed liquid crystals 133, 148
Polymers 140
Polyquinoline 15, 39
Polysiloxane 149
Polyurethanes 17
Positional disorder 136
Prepolymerization 38
Processability 21, 40
Propagation length 78
PS 144
PTCB 144
P-THEA 144
Pump-probe experiment 170
Push-pull molecules 108
PVB 144
PVK 142

Q-factor 129
Quasi-phase-matching 74
Quinoidal 112

Recombination coefficient 125
Reconfigurable interconnects 149
Refractive index 78, 94, 95
Relaxation process 70
Resolution 147
Response time 145
Reverse saturable absorption 174
Rhodamine 6G 147

Second harmonic generation 9, 52, 103
Second-order NLO properties 6
Second-order nonlinear polarization 103
Semiconductor multiquantum wells 120
Semiconductor quantum dots 147
Sensitizer 140,143
Shelf lifetimes 134
Side-chain liquid crystalline polymer 56, 69
Side-chain NLO polymers 17

Side-chain polymer 132
Silica matrix 44, 47
Singel-dendron-modified NLO
    chromophores 32
Singlet oxygen 184
Site-isolation effect 35
Slab-type waveguide 21
Smetic liquid crystals 54
Sol-gel 133
Sol-gel process 44
Space-charge distribution 123
Space-charge field 123, 126
Spatial isolation 38
Spherical Bessel functions 68
Spin coating 17
Spontaneous supramolecular
    organization 29
Static field poling 50
Statistical orientation models 64
Stilbazolium salt 19
Stilbene chromophore 23
Structural flexibility 133
Supramolecular architectures 19
Susceptibility 98

Taylor's series 105
t-Butyldimethylsilyl group 34
Temporal stability 39, 47, 48
Tensor 98
Thermal generation rate 125
Thermodynamic average 115
Thick phase gratings 121
Thin grating limit 147
Thioxanthene unit 144
Third-order nonlinear optical
    susceptibility 7
Three-dimensional microfabrication 180
Three-dimensional optical data storage 178
Three-dimensional silica network 44
Tilted configuration 135
Time-average interferometry 149
Time-gated holography 151
TNF 142
TNFDM 143
Trans-cis isomerization 60, 61, 64
Transient absorption 170, 172
Transition dipole moment 110
Transport 133
Trap limited regime 128
Traps 123
Tricyanofuran (FTC) 32
Tricyanofuran-based NLO chromophore 41
Trifluorovinylether-containing
    dendrons 38

Two-beam coupling 129, 130
Two-level model 6, 13, 14, 25, 110
Two-photon absorption 159, 161, 163, 166,
    169, 171, 172, 173, 174, 178, 179
Two-photon absorption cross-section 159,
    161, 162, 163
Two-photon confocal laser scanning
    microscopy (2PCLSM) 182
Two-photon fluorescence microscopy 181
Two-photon laser scanning microscopy
    (2PLSM) 182
Two-photon photodynamic therapy 184

Ultrashort electric pulse generation 79
Uniaxial 99
Up-converted fluorescence emission 169

Up-converted lasing 176, 189
UV-cured epoxy film 45

Van der Waals component 137
Vertical asymmetric coupler 48

Wave equation 95
Waveguide 20
Wavelength 95
Wigner's rotation matrices 65

Z-scan 172
Zwitterionic 112
Zwitterionic chromophore 73
Zwitterionic molecule 71, 72
–, electrons 94

Printing (Computer to Plate): Saladruck Berlin
Binding: Stürtz AG, Würzburg

T